TM 9 I0060542

PLUS SUPPLEMENTAL MATERIAL FROM
TM 9-1005-222-35 and FM 23-5
Department of the Army Technical Manual

RIFLE, CALIBER .30
M-1 GARAND
TECHNICAL MANUAL

OPERATOR AND ORGANIZATIONAL
MAINTENANCE MANUAL INCLUDING
REPAIR PARTS AND SPECIAL TOOLS LIST

RIFLE, CALIBER .30 M 1
RIFLE, CALIBER .30 M 1C (Sniper's)
and
RIFLE, CALIBER .30 M 1D (Sniper's)

Ⓔ Ⓟ Ⓑ Ⓜ
ECHO POINT BOOKS & MEDIA, LLC

HEADQUARTERS, DEPARTMENT OF THE ARMY, 17 March, 1969

Published 2015 by Echo Point Publishing, Brattleboro, Vermont
www.EchoPointBooks.com

TM 9-1005-222-12 Rifle, Caliber .30, M-1 Garand
ISBN: 978-1-62654-330-0 (casebound)
978-1-62654-329-4 (paperback)
978-1-62654-331-7 (spiral bound)

Cover design by Rachel Boothby Gualco
Editorial and proofreading assistance by Ian Straus,
Echo Point Books & Media

Printed and bound in the United States of America

*TM 9-1005-222-12
TECHNICAL MANUAL · HEADQUARTERS, DEPARTMENT OF THE ARMY
No. 9-1005-222-12 Washington, D.C., 17, March 1969

Operator and Organizational Maintenance Manual

RIFLE, CALIBER .30: M 1, M 1C (Sniper's), M 1D (Sniper's)

Editor; Jeff Lesemann

Assistant Editor; Francois Rheault   Artist; Alan Longwith

This manual is current as of 2 December 1968

*This manual supersedes TM 9-1005-222-12P/2, 11 August 1965 in its entirety.

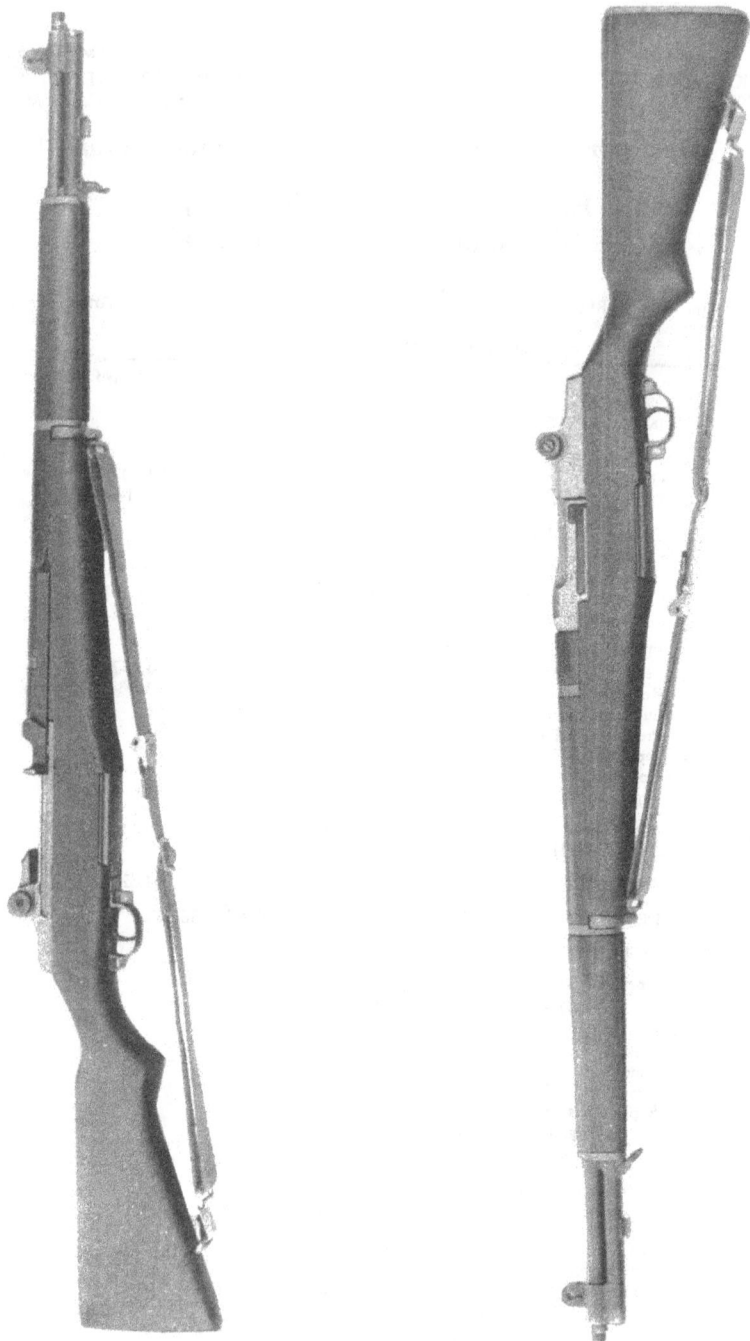

Figure 1. U.S. Rifle, Caliber .30, M 1

PREFACE

This book is the second in our series of small arms technical manuals, and it represents something of a departure from the original U.S. Army publications on which it is based. We have added several photos and sequences which were not shown in any of the Army manuals, and we have used different camera angles in many other photos, in order to provide a better view of the operations being performed. It is hoped that these changes will be viewed as improvements upon the original texts.

We have also strayed rather far afield in writing the introductory chapter on the history of the M 1. Since the available space would not permit a detailed history of the rifle, nor an adequate biography of John C. Garand, himself, the chapter focuses on the long and tortuous development process which led to its adoption. It is hoped that the chapter will lead to an appreciation of the background of this remarkable weapon and its important place in military history.

We would also like to express our thanks to Springfield Armory, Inc., of Geneseo, Illinois, for providing information and the photo of their version of the T 26 "Tanker" Garand, surely one of the most legendary side lights to the M 1 story.

The real "stars" of the book, however, are the rare and fascinating weapons which we were able to photograph, thanks to three of the finest and friendliest gun collectors one could ever hope to meet. Bill Douglas, of Dunedin, Florida,would never part with his valuable M 1C. Just imagine letting someone take a screwdriver to it, for the sake of a couple of pictures! Pierre Posse, of Sebring, Florida, trekked home, some sixty miles, to fetch his Pedersen rifle, just so we could photograph it at a gun show! Finally, Ronnie Butler, of Lakeland, Florida, invited us to his home twice to examine and photograph several prized pieces from his collection, including the extremely rare Mondragon rifle. Not only does Ronnie have one of the finest collections of military rifles around, but his charming wife, Katrinka, shares in his hobby with interest and enthusiasm to match his own. Some guys just have it all!

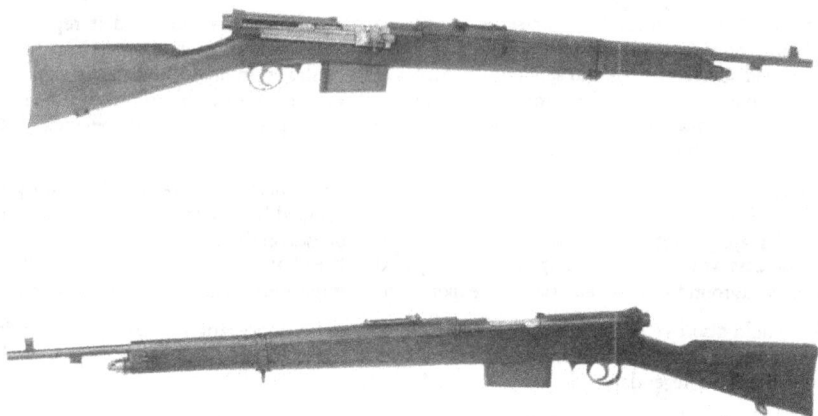

*Figure 2. Mondragon semi-automatic rifle.
From the collection of Ronnie Butler.*

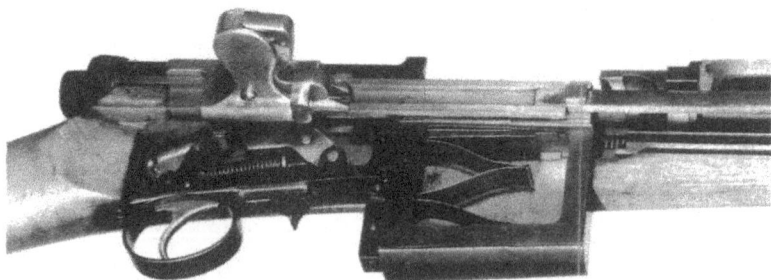

SIG 760 550

*Figure 3. S.I.G. Factory cut-away photo of
Mondragon rifle. Courtesy of Ronnie Butler.*

Introductory Chapter
"Origins and Development of the M 1 Garand"
By Jeff Lesemann

"The M 1 rifle is the greatest battle instrument ever devised"

General George S. Patton

On first glance, Patton seems to have been making another of the sweeping exaggerations which characterized so many of his public statements. But was it really such an exaggeration? In order to properly understand the M 1 rifle and the reputation that it earned, we ought to examine the complex story of its development. We must also remember the context in which General Patton's high praise was given. Only then can the relative merits and drawbacks of the M 1 Garand be accurately weighed.

The semi-automatic military rifle had its origins in one of the most unlikely places imaginable. General Manuel Mondragon, of the Mexican Army, was something of a ballistics expert and the inventor of a straight-pull bolt action rifle, in which the bolt handle cammed a rotating bolt face to unlock the breech and open the acton. These rifles, which are now extremely rare, were chambered in an equally rare 5 mm caliber. while General Mondragon was serving as Mexico's Military Attache' to France, he developed his bolt action rifle into a practical, gas-operated, semi-automatic weapon. In 1907 he patented the rifle in the United States, and he then had it built by S.I.G., in Switzerland. The Mexican Army accepted the rifle for service and placed an initial order for 4,000 pieces. Production began in 1912, but the outbreak of World War I interrupted delivery after only a handful of rifles had reached Mexico. Instead, they went to Germany, where they were issued to the Flying Corps. General Mondragon was caught up in the turmoil of Mexico's revolution of 1916, after which he and his remarkable rifle both faded into obscurity.

The Mondragon semi-automatic rifle (see fig. 2) was a remarkable and well made weapon, with a number óf advanced features, including a gas cut-off, which had the effect of converting the rifle back to a straight-pull bolt action, and a bolt disconnector, enabling the bolt to be opened manually for loading and cleaning. Some models of the rifle also featured a detachable box magazine, with a capacity of up to twenty rounds. It was chambered for the excellent 7 mm Mauser cartridge, and the bolt was designed with four extremely strong locking lugs. (see fig.3) Considering the difficulties which were to plague the development of a successful semi-automatic military rifle, this obscure pioneer showed truly remarkable sophistication!

The next semi-automatic rifle to be developed, and the first to be presented to the U.S. Army for trials, came from a Danish inventor, with the extraordinarily appropriate name of Soren Bang. His rifle was submitted for tests at Springfield Armory in 1911, and it showed considerable promise. It was another gas-operated design, with a method of operation that harkened back to some of the early experiments of John Browning. On Bang's rifle, a funnel-like cap was fitted over the muzzle. As the bullet passed by, the gasses behind it filled the cap and pushed it forward, operating the action through linking arms. It all worked quite well, but the rifle suffered from severe overheating problems, which were never solved. In 1927 Bang submitted another rifle, but it, too, was unsuccessful. Mr. Bang was not heard from again.

The French developed a semi-automatic rifle in 1916, which was built at the St. Etienne Arsenal. (see fig. 4) Although it was built in substantial numbers and issued to French troops toward the end of World War I, this gas-operated weapon, built around the rimmed 8 mm Lebel cartridge, was truly an armorer's nightmare! (see fig. 5) Both the bolt and the magazine were prone to failure, and the St. Etienne passed from the scene following the end of World War I.

In keeping with the international flavor of this story, the next entry in the semi-automatic sweepstakes came from a Chinese inventor! T.E. Liu, of China's Hangyang Arsenal, built two slightly different gas-operated rifles, which were submitted in 1918. They used the gas-cap idea, previously tried on the Bang rifle, but they were equally unsuccessful. The problem with this

Figure 4. French St. Etienne semi-automatic rifle, model 1916. From the collection of Ronnie Butler.

Figure 5. Close-up of the St. Etienne rifle with action and magazine open. From the collection of Ronnie Butler.

system was that, by trapping the gas around the muzzle, it could not then be dissipated quickly enough to prevent overheating. In 1920, James L. Hatcher, younger brother of the renowned Major General Julian Hatcher, submitted his own version of a semi-automatic rifle using the muzzle cap. It also failed, due to the same old problem of overheating, and that was to be the end of efforts to build a gas-operated rifle, using a moveable muzzle cap.

In 1918 a Swiss weapon, based on the Schmidt-Rubin rifle, was presented. It was a recoil-operated rifle, but it was no more successful than the early gas-operated types had been. Later, the Stevens Arms Company tried to improve upon the Schmidt-Rubin, but they, too, were unable to perfect a recoil-operated weapon.

By then, with the nation in the midst of World War I, the frustrating search for a semi-automatic military rifle had drawn a great deal of publicity and interest. It was at this time that a Canadian born engineer, named John C. Garand, came forward with a totally new approach. He came up with a method of utilizing the much lower energy developed by the ignition of the primer cap in order to operate the action. In his design, the firing pin was connected to a piston, which, in turn, was fitted inside the hollow bolt, locking it in place when when the action was closed. When the rifle was fired, the slight reaction of the primer cap would cause the pin to rebound, freeing the piston and allowing the bolt to open. Garand's rifle was built in 1919, while he was employed as a Civil Servant in the National Bureau of Standards, and the prototype had a lot going for it. For one thing, it really worked!

Although Garand's first prototype did not work quite well enough to win full acceptance, it was so promising that the Army definitely wanted to keep John C. Garand around. He was made a civilian employee of the Army Ordnance Department, at the princely salary of $3,500 per year, and charged with the task of perfecting his rifle. It was to be a formidable challenge, indeed, and one which would not be fully completed for another twenty years.

While John Garand was at work, others continued to submit designs of their own. Marcellus H. Thompson, who was soon to develop the famous Thompson Submachine Gun, came up with a rifle that worked on a retarded blow-back principle. Although it was a good looking weapon, its action was too violent, due to the fact that the bolt opened while the bullet was still in the barrel and under tremendous pressure. The Thompson also required specially lubricated ammunition, without which it would not function, at all.

The Thompson may have been a good looking rifle, but surely the ugliest weapon ever submitted was the French Berthier. It was a grossly overweight monstrosity, with its magazine mounted on top of the weapon, squarely in the shooter's line of sight. It didn't really matter, because the weapon worked about as well as it looked.

Other American inventors were also hard at work. Engineering teams at both Springfield Armory and Rock Island Arsenal came up with prototypes, based on gas operated conversions of the M 1903 Springfield rifle. Meanwhile, Thompson made several more efforts to refine his design. None of these weapons were fully successful, however, and toward the mid 1920's a general sense of frustration seems to have set in. Even Garand, who had built a second primer operated rifle, was no more successful than anyone else.

It began to appear that the problem was with the ammunition. The .30-06 cartridge, so named because it was adopted in 1906 for the M 1903 Springfield rifle, was arguably the finest center-fire rifle ammunition ever developed. As a military round, it was tremendously effective, and it survives to this day in a host of sporting applications which can be used against any target from varmints to big game. Its internal ballistics, however, are horrendous! At the moment of detonation, the cartridge generates a pressure of more than 50,000 pounds per square inch, high enough to have damaged receivers of some of the early Springfields. The problems with the Springfields were solved with improved metallurgical techniques, but things were not that simple for the inventors of a semi-automatic rifle using the same cartridge. While the bolt action Springfield could be strengthened to handle the load, the semi-automatic rifles had to strike a three way compromise between considerations of strength, function, and weight. Perhaps it couldn't be done.

One inventor had come up with a very successful semi-automatic weapon. In 1918, J.D. Pedersen developed a truly unique device with which the M 1903 Springfield could be converted into a semi-automatic weapon, capable of firing specially designed .30 caliber pistol cartridges from a forty round magazine. Having achieved success with this design, Pedersen entered the quest for a semi-automatic rifle by persuading the Army to consider one based on a smaller caliber cartridge. He then proceeded to develop just such a weapon.

Pedersen even cooked up the ammunition for his proposed new rifle. He selected a .276 caliber, somewhat similar to 7 mm Mauser, but with a slightly lighter powder charge, pushing a 125 grain boat-tailed bullet. Frankford Arsenal prepared a batch of the ammo and ran chronograph tests through a specially made barrel. The tests showed a muzzle velocity of 2,700 feet per second and an energy level just below 2,000 foot pounds. The round also demonstrated a flatter mid range trajectory than the .30-06, all with substantially lower internal pressures. It looked as though Pedersen might really be on to something.

The rifle which Pederson developed for his new ammunition showed even more promise. It was operated on a retarded blow-back principle, as in the Thompson, but with a toggle action mechanism which closely resembled that of the Luger pistol. (see figs. 6 & 7) When the Pedersen was submitted for preliminary trials, in 1926, the rifle and its ammuniton both performed better than any previous contender.

By using a combination of new ideas and proven technology from previous weapons, Pedersen seemed to have solved many of the problems which had plagued the semi-automatic rifle program. For instance, his ammunition had to be lubricated, like that used in the Thompson rifles, but Pedersen's method of lubricating the cartridges was neat and virtually undetectable. He had the cartridges dipped in a solution of solvent and mineral wax. When the solvent evaporated

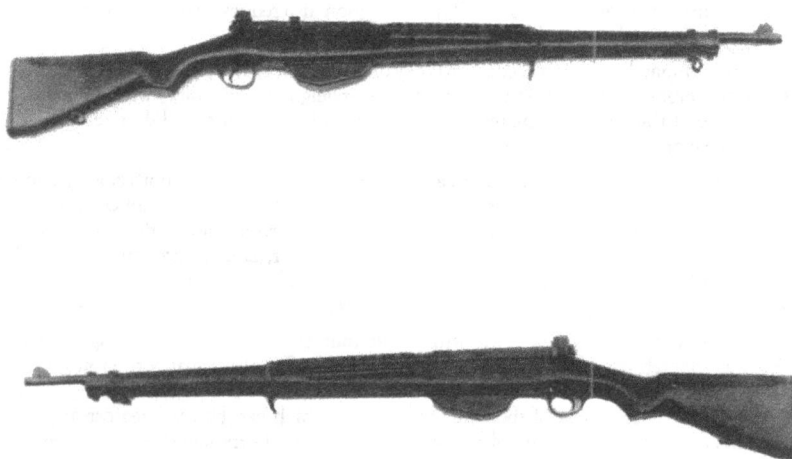

Figure 6. Pedersen .276 caliber semi-automatic rifle. From the collection of Pierre Posse.

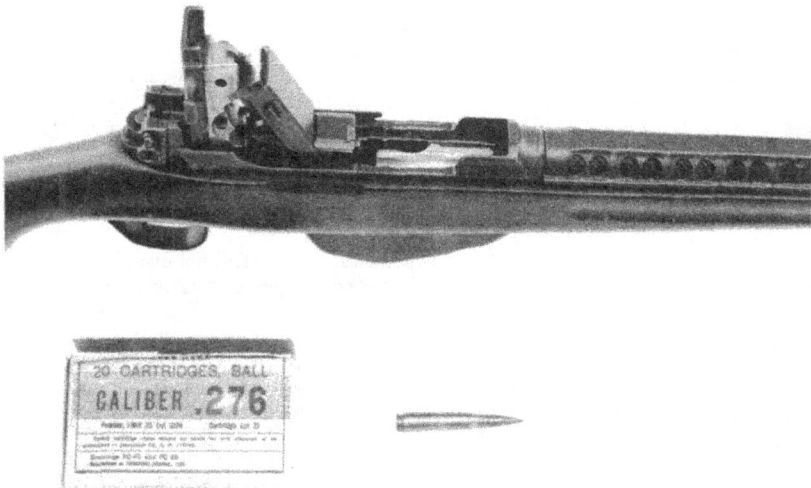

Figure 7. Close-up of Pedersen rifle - action open. Together with cartridge and box of ammunition. From the collection of Pierre Posse.

it left behind a thin, hard coat of dry wax that was neither sticky nor slippery to the touch. When a round was fired, however, the wax would melt instantly, providing a slick surface for easy extraction. The cartridges were loaded into ten round expendable clips, an idea borrowed from the old Mannlicher rifles. The clip was loaded, en bloc, into the magazine, where it served as an integral component of the feeding system. When the last round in the clip was fired, it would be ejected from the rifle, along with the spent case. Despite the fact that there were some early problems during tests, in which the clip would be ejected prematurely, the system worked well enough to show promise, and many key officers in the Ordnance Department favored the use of such clips. In fact, a great many of them began to favor the adopton of the Pedersen, period!

While Pedersen was apparently on the brink of success, Garand's fortunes could hardly have been worse. In 1926 the .30-06 cartridge was modified, in order to give it even more power than before. The new load was powerful enough to completely eliminate the possibility of perfecting a primer actuated semi-automatic rifle. All of Garand's work on such a weapon was thus rendered useless. In an attempt to mount some sort of a challenge to the Pedersen, several officers of the Ordnance Department persudaded Garand to build his own .276 rifle, and, since the primer actuated mechanism seemed to be a dead end, he decided to give the gas operated principle one more try. It was this fortunate turn of events, which led to Garand's final triumph. After all, we are telling the story of the M 1 Garand, and not the "M 1 Pedersen!"

The breakthrough centered around the method by which the gas was channeled and utilized. Instead of using a moveable gas funnel, as all of the unsuccessful prototypes had done, Garand channeled the gases into a cylinder under the barrel, where its pressure would act upon a piston. The piston, in turn, operated the bolt. All of this was accomplished without excessive heat build-up, because the gases could continue to expand freely.

IT WORKED! Garand's .276 Pedersen prototype was the first rifle which looked like the Garand that we know today, and it succeeded in beating several of the Pedersen's own best features. For one thing, Garand developed a flexible magazine follower assembly, which simplified loading. The Pedersen clip had a definite top and bottom to it, and cartridges had to be loaded into the clip in one precise way, or they would not feed. Garand's clip was a simple stamping, designed in such a way that it could be inserted with either end up. Also, the cartridges could be loaded with the top and bottom cartridges on either the right or the left side of the clip. If a weapon must have an en bloc clip, this is clearly the best way to do it. What's more, the gas operated Garand rifle did not require specially lubricated cartridges. In 1929 formal tests were held, and the new Garand rifle clearly beat the Pedersen.

The long process of developing a new semi-automatic rifle was still not over, however. With the onset of the Great Depression, the meager budgets of the peacetime Army shriveled almost to nothing. In addition, the question of the caliber was still not resolved, and the Ordnance Department decided to use what little resources they had in an effort to compare .276 Pedersen and .30-06. A committee was established and tests were conducted by firing the two calibers into animal carcasses at various ranges. At close ranges the .276 bullet made a very nasty wound, but its accuracy and effectiveness both deteriorated as the range increased. The conclusion favored .30-06, and the two contenders were directed to develop their rifles in the larger caliber.

By now, the pressure was on Pedersen. The success of the .276 Garand was rather easily translated into the larger caliber, but the retarded blow-back, toggle action Pedersen simply could not handle the higher power ammunition. While Garand was now put to work developing the M 1 for production, Pedersen tried unsuccessfully to sell his rifle abroad. He built a small run of rifles for trial in England, but that was as far as he got. Ironically, the Japanese built a copy of the Pedersen during World War II. It seems that 7mm Arisaka was very close to .276 Pedersen. While the Japanese version of the rifle seems to have worked satisfactorily, Japan was also unable to develop it. The few surviving examples of the Pedersen, from any of the three countries in which it was tried, show that it came close to success, but it just did not make the grade.

In March of 1932 Springfield Armory was given an order for eighty semi-production Garands, under the designation, "U.S. Semi-automatic Rifle, Caliber .30 TIE2." This designation was officially changed on August 3, 1933 to "U.S. Rifle, Caliber .30 M 1". The use of the Garand name was always unofficial.

Over the next five years several small batches of M 1 rifles were made, using steadily improved production techniques. A few nagging problems were encountered along the way, and refinements were made to the operating cam, the clip, and the rear sight. A major change was also made to the gas cylinder, which involved the replacement of the open gas cap with the gas cylinder that incorporates the front sight, as seen on all subsequent production rifles. The gas port, bored through the barrel at the muzzle, allows gas to vent into the cylinder with sufficient velocity to prevent the build-up of powder residue. The first production M 1 rifles were finally delivered in 1937, and two years later, on September 1, 1939, production reached 100 rifles per day, just as Hitler's forces launched the invasion of Poland. The United States was destined to be the only nation on either side to have a standard production semi-automatic infantry rifle throughout all of World War II. In the light of this simple historical fact, Patton's comment on the M 1 does not seem so far from the mark, after all.

This is not to say that the M 1 did not have its shortcomings. Nor should one overlook the serious, but belated challenge mounted by the development of the Johnson rifle. (see fig.   )
The Johnson was technically significant, if only for the fact that it represented a successful attempt to design and build a working, recoil operated, semi-automatic rifle in .30-06 caliber. However, the Johnson would have been more complicated to manufacture than the Garand, and it would have been much slower to come on line at a time when the attack on Pearl Harbor showed our forces to be ill prepared for war, anyway. Years later, the development of the M 14, incorporating its features of reduced weight, the box magazine, and selective fire, would take the basic M 1 design to its fullest level of refinement. Meanwhile, the M 1 would help to win World War II on all fronts, and it would fight effectively, again, in Korea.

Figure 8. The Johnson semi-automatic rifle. From the collection of Ronnie Butler.

During the service life of the M 1 Garand, it was produced in huge numbers, primarily by the Springfield Armory, but also by several contract manufacturers, including Winchester, Harington and Richardson, and International Harvester. There are slight variations in some of the rifle's components, such as the front and rear sights, the trigger guard, and the gas cylinder cap. In addition, changes were made involving a number of the rifle's accessories, such as the bayonet, the flash hiders used on the sniper M 1's, the grenade launchers, and the winter triggers. All of these different components are given the fullest possible treatment in the the following pages, but it is impossible, within the limited scope of this manual, to provide detailed coverage for the collector, concerning the "correct" combinations of components which should be found on any one rifle.

The National Match versions of the M 1 Garand, which were produced for target competition, starting in the 1950's, are finely tuned target rifles, having specially fitted stocks, special barrels, and other custom fitted components, for maximum accuracy. While these rifles are basicly the same weapon as the standard M 1, their special nature necessitates greater than normal care when one is performing maintenance or cleaning operations on them.

A few additional notes should be added to the story of the M 1. The first item concerns the mythical T 26, "Tanker," Garand. Two of the Garand's most enthusiastic and powerful admirers, Generals Patton and MacArthur, both suggested that a shortened, carbine version of the M 1 should be developed. MacArthur wanted such a rifle for its advantages in the house-to-house style of all out fighting which he anticipated in the invasion of Japan, while Patton wanted a rifle that would fit handily inside of his Third Army tanks. The "Tanker" handle was apparently connected to Patton's association with the idea. It is a misnomer, however. The "T" designation stands for "Trial," and it has been used for all Ordnance Department prototypes, including original M 1. In 1945 approximately 1,000 T 26's were built for trials. While they worked well enough to merit acceptance, the end of the war meant the end of any real need for a carbine version of the M 1. The small number of rifles which had been completed were largely lost, but the "Tanker" became something of a legend. A large number of fake "Tankers" were built during the 1950's, and they were usually of rather poor quality. In recent years, however, Springfield Armory, Inc. (the private company) has built its own version of the "Tanker," and it is an excellent weapon, built to the original T-26 specifications. Internally, the "Tanker" is identical to the full size M 1.

Figure 9. T 26 "Tanker" Garand. Courtesy Springfield Armory, Inc.

It should also be noted that the M 1 Garand, itself, is still being manufactured by at least two private companies, including Federal Ordnance Co. and the previously mentioned Springfield Armory, Inc. These rifles use combinations of new components and surplus military parts, and they are essentially identical to the original rifles.

The pages which follow are republished from a combination of original U.S. Army Tech Manuals and Field Manuals, covering the M 1 Garand. The information is provided for general reference, and, as with any firearm, no alteration or repair should be undertaken without consulting a trained gunsmith.

BIBLIOGRAPHICAL NOTES

Considering the importance and the abundance of the M 1 Garand, there is relatively little published material on the rifle. Among the books which are available, however, are the following:

HATCHER'S BOOK OF THE GARAND; by Major General Julian Hatcher, currently being reprinted by Gun Room Press.

HATCHER'S NOTEBOOK: by the same author, currently reprinted by Stackpole Books.

KNOW YOUR M 1 GARAND; by E.J. Hoffschmidt, published by Blacksmith.

THE BOOK OF RIFLES; by W.H.B. Smith, currently out of print, but widely available on the used book market.

SMALL ARMS OF THE WORLD; originally by W.H.B. Smith, later editions by others, including Edward C. Ezell. The 12th edition of this book is currently published by Stackpole Books.

Chapter 1
INTRODUCTION
Section I. General

1-1. Scope
These instructions are for use by the operator and organizational maintenance personnel. They apply to Caliber .30 Rifles, M1, M1C (Sniper's) and M1D (Sniper's).

1-2. Forms and Records
    a. General. Refer to TM 38-750 (Army Equipment Records Procedure) for forms and records required.
    b. Recommendations for Maintenance Manual Improvements. Report of errors, omissions, and recommendations for improving this publication by the individual user is encouraged. Reports should be submitted on DA Form 2028 (Recommended Changes to DA Publications) and forwarded direct to:
    Commanding General
    U.S. Army Weapons Command
    ATTN: AMSWE-SMM-P
    Rock Island, Illinois 61201

1-3. Administrative Storage
Refer to TM 740-90-1 for administrative storage.

Section II. DESCRIPTION AND DATA

1-4. Description
    a. General. The Rifles, M1, M1C (Sniper's) and M1D (Sniper's) (figs. 1, 10 and 11) are clip-fed, gas-operated, air-cooled, semiautomatic shoulder weapons.
    b. Differences in Models.
    (1) The M1C has a telescope mounted to the receiver.
    (2) The M1D has a telescope mounted to the barrel.
    (3) The M1C and M1D also require a flash hider and a cheek pad.

1-5. Tabulated Data
    a. Rifle, M1.
Weight of rifle w/o equipment .... 9.5 lb. approx.
Weight of rifle w/bayonet .... 10.5 lb. approx.
Length of rifle .... 43 in.
Length of barrel .... 24 in.
*Muzzle velocity .... 2,750-2,800 fps
*Maximum effective range .... 500 yd.
*Maximum effective rate of fire (aimed
    rounds per minute) 16-24
*Number of cartridges in clip 8
*Types of ammunition .... Ball, armor-piercing-incendiary,
                              tracer, blank, rifle grenade cartridge and dummy
    b. Rifles, M1C (Sniper's) and M1D (Sniper's).
Weight w/equipment (telescope, flash hider,
    gun sling, and cheek pad) .... 11.75 lb. approx.
Length of rifle w/flash hider, type T-37 .... 46-1/8 in.

*This information also applies to the M1C and M1D Rifles.

Figure 10. U.S. Rifle, Caliber .30, M 1C (Sniper's), with flash hider, M 2. From the collection of Bill Douglas.

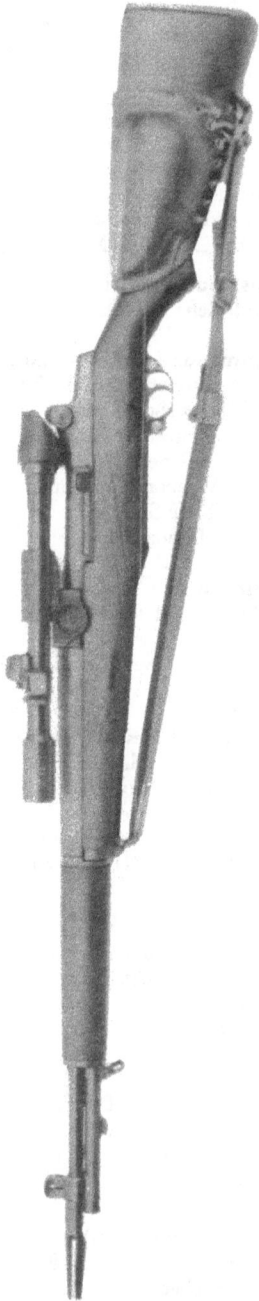

Figure 11. U.S. Rifle, Caliber .30, M 1D (Sniper's) with flash hider, T 37.

Chapter 2
OPERATING INSTRUCTIONS
Section I. Controls

2-1. General

This section describes, locates, illustrates, and furnishes the operator with essential information pertaining to the various controls provided to properly operate the materiel.

2-2. Controls

Refer to table 2-1.

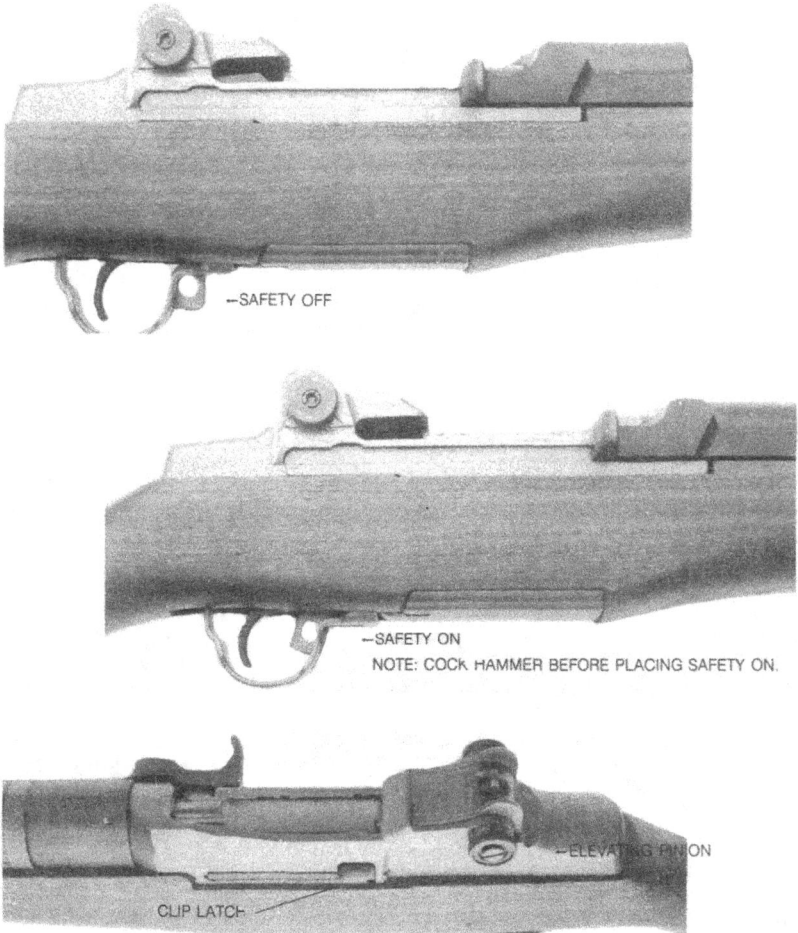

Figure 12. Controls.

Table 2-1. Controls

Item (See fig. 12)	Purpose
Safety	To prevent accidental firing.
Trigger	To release hammer to effect firing.
Windage knob	To adjust lateral movement of rear sight.
Elevating pinion	To adjust elevation of aperture.
Clip latch	To hold clip in receiver until last round is fired.

Section II. OPERATION UNDER USUAL CONDITIONS

2-3. General

This section contains instructions for the operation of the rifles under conditions of moderate temperatures and humidity. Instructions for operation under unusual conditions are covered in section IV.

2-4. Preparation for Firing

a. Examine bore. Make certain it is free of powder fouling or corrosion.
b. Check gas cylinder lock screw for secure installation.
c. Check ammunition. Make certain it is clean and that it is of the proper type and grade
d. Cock the rifle and place the safety in safe position (fig. 12).

2-5. Service Before Firing

Perform the before firing operations as indicated in table 3-3.

2-6. Loading

Refer to FM 23-5.

2-7. Zeroing

Refer to FM 23-5.

2-8. Misfire, Hangfire, and Cook-off

Refer to FM 23-5 and paragraph 2-9b, below.

2-9. Procedures for Removing a Round in Case of Failure to Fire

a. **General** After failure to fire, due to misfire, the following general precautions, as applicable, will be observed until the round has been removed from the weapon and the cause of failure determined.

(1) Keep the weapon trained on the target and see that all personnel are clear of the muzzle.

(2) Before retracting the bolt and removing the round, see that personnel, not required for operation, are cleared from vicinity.

(3) Make certain the round, removed from the weapon, is kept separate from other rounds until it has been determined whether the round or weapon is at fault. If the weapon is determined to be at fault, the round may be reloaded.

b. **Time Intervals.** The definite time intervals for waiting, after failure of weapon to fire, are prescribed as follows: Always keep the round in the chamber for five seconds from the time a misfire occurs to insure against an explosion outside of the gun in event a hang-fire develops. If the barrel is hot and a misfire stops operation of the gun, wait five seconds with the round locked in the chamber to insure against hangfire dangers (a hangfire will occur within five seconds after the primer is struck), then extract the round immediately to prevent cook-off. If the round cannot be extracted within an additional five seconds, it must remain locked in the chamber for five minutes because of the possibility of a cook-off. Also in the event the barrel is hot and misfire occurs when attempting to resume firing after an intentional cessation of firing, the round should remain locked in the chamber for five minutes because of the possibility of a cook-off.

2-10. Service During Firing

Perform the during firing operations as described in the operators preventive-maintenance services (table 3-3).

2-11. Unloading

Refer to FM 23-5.

2-12. Service After Firing

Perform the after firing operations as described in the operator's preventive-maintenance services (table 3-3).

Section III. OPERATION OF MATERIEL USED IN CONJUNCTION WITH MAJOR ITEM

2-13. General

The following materiel is not normally used continually. Therefore, it is necessary to protect from weather and dampness in storage. Clean and lubricate materiel as required, whether in use or in storage.

2-14. Equipment

a. **Grenade Launcher, M7A3 and Grenade Launcher Sight, M15.** Refer to FM 23-30.

b. **Bayonet-Knife, M5 and M5A1 and Bayonet-Knife Scabbard, M8A1.** Keep bayonet in scabbard except when removed for training, inspections, cleaning, repair, or for use in combat or danger zones.

c. **Winter Trigger Kit.** Use the winter trigger kit only in extreme cold operation and on authority of unit commander.

Section IV. OPERATION UNDER UNUSUAL CONDITIONS

2-15. General

Report any chronic failure of materiel resulting from subjection to extreme conditions (par 1-2).

2-16. Operation in Extreme Cold

a. In climates consistently below 0°F, it is necessary to prepare the rifle for cold-weather operation. The rifle should be thoroughly cleaned wiith SD, dry cleaning solvent, and lubricated with LAW, weapons lubricating oil.

b. Rifles should be free of moisture and excess oil. Moisture or too much oil on the working parts will cause them to be sluggish in operation, or perhaps to fail completely.

c. Exercise moving parts through their entire range at required intervals. This movement helps prevent parts from freezing in place and reduces effort required to operate them.

d. Materiel not in use and stored outside must be protected with proper cover.

Note. Transferring weapon from cold to warm air may cause moisture to collect. If the weapon is brought into a warm room, allow it to reach room temperature before cleaning and lubricating as required.

2-17. Operation in Extreme Heat

a. **Hot, Dry Climates.**

(1) The film of oil necessary for operation and preservation dissipates quickly in hot climates. Inspect the rifle, paying particular attention to all hidden surfaces such as bolt and lug, operating rod and recess, cam surfaces and bolt locking recess in receiver, where corrosion might occur and not be quickly noticed.

(2) Perspiration from the hands contributes to rusting because it contains acids and salts. After handling materiel, clean, wipe dry, and restore the oil film using PL special, general purpose lubricating oil.

(3) Clean and oil the bore more frequently than usual.

(4) Apply linseed oil to wooden parts to prevent drying.

b. Hot, Damp, and Salty Atmosphere.

(1) See a(1) and (2) above under hot, dry climates.

(2) Inspect materiel frequently because of increased possibility of rust.

(3) When material is active, clean and lubricate the bore and exposed metal surfaces more frequently than prescribed for normal service.

(4) Moist and salty atmosphere tend to emulsify oils and greases and destroy their rust-preventive qualities. Inspect all parts frequently for corrosion.

(5) When materiel is inactive, cover metal surfaces with a film of PL special, general purpose lubricating oil.

(6) Apply linseed oil to wooden parts to keep out moisture.

2-18. Operation in Dusty or Sandy Areas

a. Clean and lubricate the materiel more frequently in sandy or dusty areas. Exercise particular care to keep sand out of mechanisms when carrying out inspecting and lubricating operations. Shield parts from flying sand with tarpaulins during disassembly and assembly operations.

b. Before operating in sandy areas, remove lubricant from bolt, barrel and receiver, operating rod, and trigger housing assembly, as they will pick up sand and from an abrasive which will cause rapid wear. Dry surfaces wear less than surfaces coated with lubricants contaminated with sand. Clean and lubricate all exposed parts after action is over.

2-19. Hand-Carried Fording

a. No special lubricaion is required before fording.

b. Protect from water splashes.

c. If immersion does occur, proceed as directed in paragraph 2-20.

2-20. Maintenance After Immersion

a. General. During hand-carried fording, water seepage into bolt, trigger housing, receiver, and operating rod assembly will usually occur. It is advisable, therefore, that the service outlined below be accomplished on all weapons submerged in water as soon as practical to prevent damage to the weapon.

b. Procedures.

(1) After submersion in salt water, wash in clear water to remove corrosive salts.

(2) Drain all trapped moisture and wipe dry.

(3) Assemblies which require disassembly for proper lubrication must be disassembled, dried, and lubricated as soon as possible.

Note. Items not authorized for operator disassembly/assembly must be cleaned by organizational maintenance personnel.

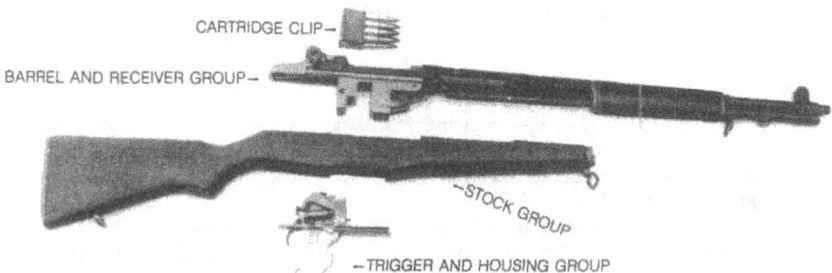

CARTRIDGE CLIP—

BARREL AND RECEIVER GROUP—

—STOCK GROUP

—TRIGGER AND HOUSING GROUP

Figure 13. U.S. Rifle, Caliber .30, M 1. Major assembly group.

Chapter 3
SERVICE AND MAINTENANCE INSTRUCTIONS
Section I. Service Upon Receipt of Materiel

3-1. General

a. When a new or reconditioned Rifle, M1, M1C (Sniper's) or M1D (Sniper's) is received, it is the responsibility of the officer in charge to determine whether the materiel has been properly prepared for service by the supplying organization and to be sure it is in condition to perform its function.

b. All basic issue items will be checked with the listing in appendix B.

c. A record will be made of all missing parts, tools, and equipment, and of any malfunctions. Corrective action should be initiated as soon as possible.

3-2. Services

Refer to table 3-1 for services performed on receipt of materiel.

Table 3-1. Service upon Receipt of materiel

Step	Action	Reference
	RIFLES	
	Note. When new rifles are received, they are sealed in vapor proof, volatile corrosion inhibitor (VCI) bags. They are packed two in a carton and five cartons in a box.	
1	Remove carton from box and rifle from carton and bags.	
2	Check for missing items. *Note.* Items must agree with Basic issue items list.	App B, Sec II
3	Clean and lubricate bore and chamber.	Par 3-11a and 3-6
4	Field strip and inspect for missing parts and proper assembly.	FM 23-5
5	Clean and lubricate the following: Locking lugs of bolt Bolt guides Camming surfaces of operating rod	Par 3-11a and 3-6
6	Perform "before operation" preventive maintenance checks and services.	Table 3-3

Section II. REPAIR PARTS, SPECIAL TOOLS AND EQUIPMENT

3-3. Tools and Equipment

Tools and equipment issued with the Caliber .30 Rifles, M1, M1C (Sniper's) and M1D (Sniper's) are listed in the basic issue items list, appendix B.

3-4. Special Tools and Equipment

Special tools and equipment are listed and illustrated in appendix B.

3-5. Maintenance Repair Parts

Organizational maintenance repair parts are listed and illustrated in appendix B.

Section III. LUBRICATION INSTRUCTIONS

3-6. General

a. Make certain all metal parts have been cleaned with SD, dry cleaning solvent. Dry thoroughly. Apply a light coat of preservative, PL special, general purpose lubricating oil, for above 0°F, and LAW, weapons lubricating oil, for below 0°F. Apply a light coat of rifle grease to the following surfaces:

(1) Locking lugs of bolt, operating lug, and recesses.

(2) Bolt guide.

(3) Cams on trigger and hammer.

b. Refer to table 3-2 for a listing of lubrication and cleaning materiels and stock numbers for requisitioning purposes.

c. Refer to paragraph 2-16 thru 2-20 for specific lubrication instructions under unusual conditions.

Table 3-2. materiels Required for Maintenance Functions

Federal stock number	Item
8020-244-0153	BRUSH, ARTISTS: metal, ferrule, flat, chisel edges, 7/16 lg. exposed bristle
7920-295-2491	BRUSH, CLEANING, TOOL AND PARTS: rd, 100 percent tampico fiber.
6850-965-2332	CARBON REMOVING COMPOUND: (P-C-111) (5 gal. pail)
	CLEANING COMPOUND, RIFLE BORE: (CR)
6850-224-6656	2 oz. can
6850-224-6657	6 oz. can
6850-224-6658	1 qt. can
5350-221-0872	CLOTH, ABRASIVE: crocus, ferric oxide and quartz, jean-cloth-backing, closed-coating. (CA)
6850-281-1985	DRY CLEANING SOLVENT: (SD) (1 gal can)
	LUBRICATING OIL, GENERAL PURPOSE: (PL special)
9150-273-2389	4 oz. can
9150-231-6689	1 qt. can
9150-754-0063	GREASE, RIFLE: (1 lb. can)
8010-221-0611	LINSEED OIL, RAW: (1 gal. can) (TT-L-00215)
9150-292-9689	LUBRICATING OIL, WEAPONS: (LAW) for below zero operations (1 qt. can)
7920-205-1711	RAG, WIPING: cotton (50 lb. bale)

Section IV. PREVENTIVE MAINTENANCE CHECKS AND SERVICES

3-7. Preventive Maintenance

a. Purpose. To assure maximum operational readiness the operator must perform certain scheduled maintenance services at designated intervals. See basic preventive maintenance procedures (1) thru (3) under b below, and table 3-3.

b. Preventive Maintenance Performed by Operator.

(1) Rust, dirt, grit, gummed oil, and water cause rapid deterioration of outer surfaces and internal mechanisms. Exercise care to keep all surfaces clean and properly lubricated. Exterior surfaces of the weapon (or components) are not to be cleaned or polished with treated cloths or other commercial compounds.

(2) Tighten loose parts.

(3) Every six months check to see if all modifications have been applied. Refer to DA Pam 310-7. No alteration or modification will be made except as authorized by modification work order.

Table 3-3. Preventive maintenance Checks and Services

Item Number	Interval Operator Daily B	D	A	Org. W	Item to be inspected	Procedure	Reference
					B—Before operation D—During operation	A—After operation W—Weekly	
1	1	—	—	—	M1, M1C, M1D	Hand cycle the action to insure binding is not present.	
2	2	—	—	—	Trigger housing assembly	Actuate safety. Safety will not engage when hammer is forward.	Fig. 12
3	3	—	—	—	Barrel and receiver	Actuate windage knob and elevating pinion of rear sight group for proper operation. Aperture must retain position against thumb pressure.	Fig. 12
4	4	—	—	—	Barrel and receiver	Check front sight for secure installation.	
5	—	5	—	—	M1, M1C, M1D	Check gas cylinder lock screw for secure installation. **Note.** Do not tighten lock screw when weapon is hot.	
6	—	—	6	—	M1, M1C, M1D	Clean chamber, bore, and all components.	Par 3-11a
7	—	—	7	—	M1, M1C, M1D	Lubricate.	Par 3-6

Section V. TROUBLESHOOTING

3-8. General
Refer to table 3-4 for troubleshooting.

Table 3-4. Troubleshooting

MALFUNCTION	PROBABLE CAUSE	CORRECTIVE ACTION	
		OPERATOR	ORGANIZATIONAL
Failure to load	Damaged clip	Replace clip.	
	Improperly assembled receiver components	Disassemble, and reassemble correctly (refer to FM 23-5).	
Failure to feed	Weak or broken operating rod spring	―――――――――	Replace spring (2, fig. 36).
	Binding or damaged operating rod	―――――――――	Evacuate to direct support maintenance personnel.
Bolt fails to close	Dirty or deformed ammunition	Clean or replace ammunition.	
	Cartridge case holding bolt out of battery	Pull bolt to the rear and remove dirty or deformed cartridge.	
	Dirty chamber	Clean chamber (par 3-11a).	
	Extractor does not snap over rim of cartridge	Clean bolt assembly and extractor recess (par 3-11a).	Replace extractor (5, fig. 36).
	Frozen ejector spring and plunger	―――――――――	Replace ejector (6, fig. 36).
	Restricted movement of or damaged operating rod	―――――――――	Evacuate to direct support maintenance personnel.
	Weak or broken operating rod spring	―――――――――	Replace spring (2, fig. 36).
	Damaged receiver	―――――――――	Evacuate to direct support maintenance personnel.
Failure to fire	Bolt not in battery	See "bolt fails to close".	
	Defective ammunition	Follow procedures for misfires (refer to FM 23-5).	
	Firing pin worn, damaged, or movement restricted	―――――――――	Replace firing pin (8, fig. 36).
	Inadequate firing pin protrusion	―――――――――	Evacuate to direct support maintenance personnel.
	Weak or broken hammer spring	―――――――――	Replace hammer spring (4, fig. 35).

Malfunction	Probable cause	Corrective action
Short recoil	Hammer damaged or broken	Evacuate to direct support maintenance personnel
	Gas cylinder lock screw loose	Tighten lock screw (15, fig. 36).
	Gas cylinder lock not fully seated	Tighten gas cylinder lock (16, fig. 36).
	Carbon or foreign matter in gas cylinder or barrel port	Clean (par 3-11a).
	Defective operating rod spring	Replace spring (2 fig. 36).
	Gas cylinder not fully seated	Remove and properly install gas cylinder (figs. 22, 23)
	Improper lubrication in cold weather	Clean and lubricate properly (par 2-16).
	Unserviceable gas cylinder	Evacuate to direct support maintenance personnel.
	Unserviceable gas piston	Evacuate to direct support maintenance personnel.
Failure to extract	Cartridge seized in chamber	Remove cartridge and clean chamber (par 3-11a).
	Damaged or deformed extractor	Replace extractor (5, fig. 36).
	Ruptured cartridge case	Remove cartridge case fig. 14.
Failure to eject	Short recoil	See "short recoil".
	Weak, frozen or distorted ejector spring and plunger	Replace ejector (6, fig. 36).
Failure of bolt to be held rearward after last round is fired	Insufficient rearward movement of bolt	See "short recoil".
	Defective or broken operating rod catch	Evacuate to direct support maintenance personnel.
	Defective operating rod.	Evacuate to direct support maintenance personnel.
Failure of clip to eject	Broken or deformed latch spring	Replace latch spring (27, fig. 36).

EXTRACTOR

Figure 14. Removal of ruptured cartridge case.

Section VI. OPERATORS MAINTENANCE PROCEDURES

3-9. Removal/Installation of Major Groups and Assemblies
Refer to FM 23-5. Also see fig. 13 and fig. 15.

3-10. Disassembly/Assembly Barrel and Receiver Group
Refer to FM 23-5. Also see figs. 20 through 30.

3-11. Cleaning, Inspection, and Repair
  a. Cleaning.
    (1) General.
    (a) Immediately after firing, thoroughly clean bore with a bore brush saturated with CR, rifle bore cleaning compound.

    (b) After cleaning with CR, run dry swabs thru the bore until the swabs are clean. Make certain that no trace of burned powder or other foreign substances are left in bore. Then apply a light coat of PL special, general purpose lubricating oil.

    (c) Clean the chamber with a cleaning brush dipped in CR

    (d) Clean all surfaces exposed to powder fouling (bolt face, chamber, piston area of operating rod assembly and gas cylinder lock screw) with CR.

    Note. This compound is not a lubricant. Wipe dry and oil all parts which require lubrication.

    CAUTION. The use of abrasives, steel wool, wire brushes, or scrapers on the piston area of the operating rod assembly will change critical dimensions that may cause the weapon to malfunction and is therefore prohibited. The application of lubricants to this area is also prohibited.

    (e) For general usage, SD, dry cleaning solvent, may be used to clean or wash grease and oil from all parts of the rifle.
    (2) General precautions in cleaning.
    (a) SD, dry cleaning solvent, is flammable and should not be used near an open flame. Have fire extinguishers available when using this material. This solvent evaporates quickly and has a drying effect on the skin. If used without gloves it may cause cracks in the skin; in some individuals, mild irritation or inflammation may bedevelop. Use only in well-ventiilated areas.

    (b) The use of gasoline, kerosene, bezene (benzol) or high-pressure water, steam, or air, for cleaning the weapon is prohibited.

    (c) Do not dilute CR, rifle bore cleaning compound. Do not add antifreeze. Store cleaner in a warm place. Shake CR well before using.

    (3) Cleaning of sling and scabbard. Clean mildewed canvas by scrubbing with a dry brush. If water is necessary to remove dirt, it must not be used until mildew has been removed. Oil and grease may be removed by scrubbing with issue soap and water. Rinse well with water and dry.

    CAUTION. At no time is gasoline or any solvent to be used to remove oil or grease from canvas.

To prevent mildew, air canvas items frequently.

(4) Cleaning. Clean with a dry cloth. Periodically rub raw linseed oil on wooden components to prevent drying or the absorption of moisture.

CAUTION. Do not apply linseed oil to those surfaces next to the barrel. Application of oil to these surfaces creates heavy smoking when the barrel is hot. This smoke will obscure the operator's vision. Portions which swell due to high moisture content should be dried before applying the linseed oil. Do not allow linseed oil to contact or remain on metal parts.

b. Inspection. Refer to paragraph 3-7.

c. Repair. Turn rifle into organizational maintenance personnel for any necessary repair.

Section VII. ORGANIZATIONAL MAINTENANCE PROCEDURES

3-12. Removal/Installation of Major Groups and Assemblies
Refer to FM 23-5.

3-13. Disassembly/Assembly of Major Group and Assemblies
Refer to figures 15 through 30, 35, 36 and 41.

NOTE. White dots indicate disassembly, and black dots indicate assembly.

3-14. Cleaning, Inspection, and Repair
a. Cleaning.
(1) General. Refer to paragraph 3-11a for general cleaning procedures.

(2) Removing carbon. On component parts which have a hard carbon residue it may be necessary to clean these parts with P-C-111, carbon removing compound. Observe the following procedures when using P-C-111.

WARNING. Avoid contact of P-C-111 with skin. If contact does occur, wash compound off thoroughly with running water. A good lanolin base cream is helpful if applied after washing off compound. Recommend use of gloves and protective equipment.

(a) Using a suitable container, fill with fresh compound.

(b) Before soaking a component in compound, remove all grease, dirt and oil as indicated in paragraph 3-11a. Place parts to be cleaned in a container and make certain they are completely immersed.

(c) Soak for 2 to 16 hours as necessary. Remove parts and drain. Rinse with water or solvent. To effectively remove carbon, brush with a stiff bristle brush (not wire) under running water.

(d) Wipe parts dry and lubricate (par 3-6).

b. Inspection and Repair.
Refer to table 3-5.

NOTE. For items not authorized at organizational maintenance level, evacuate to direct support maintenance personnel.

Table 3-5. Organizational Maintenance Functions

Warning: Before starting an inspection, be sure to clean the weapon. Do not actuate the trigger until the weapon has been cleared. Inspect the chamber to insure that it is empty, and check to see that no ammunition is in position to be introduced.

ITEM	INSPECTION AND REPAIR
Trigger housing assembly	Inspect for and remove burs.
	Replace items 1, 2, 4, 6, 8, and 10, Fig. 35, if worn or damaged.
Stock assembly	Inspect for cracks, breakage, or damage that would weaken the stock. Evacuate to direct support maintenance personnel.
	Check for dry, unoiled areas of wood. Oil with raw linseed oil only. Do not oil inside of stock.
	Make certain that the butt plate assembly is secure to the stock.
Barrel and receiver group	Inspect barrel for rust or obstructions in bore and remove.
	Inspect for and remove burs.
	Inspect chamber for ruptured cartridge. Remove with the ruptured cartridge case extractor.
Operating rod assembly group	Inspect for damage that may restrict movement of operating rod assembly.
	Replace item 2, fig 36·, if weak, broken, or kinked.
Bolt assembly	Inspect for and remove burs.
	Inspect firing pin. If chips or cracks are present in tip area, or if badly worn, replace.
	Inspect ejector and spring assembly hole for distortion, burs or rust that would hinder free movement of the ejector assembly.
	Replace items 5, 6, 7, and 8, fig 36, if worn or damaged.
Follower group	Replace item 10, fig 36, if worn or damaged.
Gas cylinder group	Inspect front sight. make sure it is secure.
	Inspect gas cylinder lock screw. Make sure it is tight, but not "frozen" or cross-threaded in gas cylinder.
	Replace items 15, 18, and 19 fig 36, if worn or damaged.
Handguard group	Inspect for cracks that would impair serviceability. Replace items 20, 21, and 22, fig 36, if damaged.
	Note. Evacuate damaged handguards to direct support maintenance personnel for repair.
	Inspect for damaged parts.
	Replace item 27, fig 36, if worn or damaged.
Rear sight group	Inspect for damaged parts.
	Inspect windage knob and pinion for binding. Replace items 28, 29, and 30, fig 36, if worn or damaged.
	Raise the aperture to full height and reduce by four clicks. Grasp the rifle at the small of the stock with the thumb on the aperture. Press down on aperture. Aperture should not move under thumb pressure.
Assembled rifle	Hand operate to assure proper functioning.

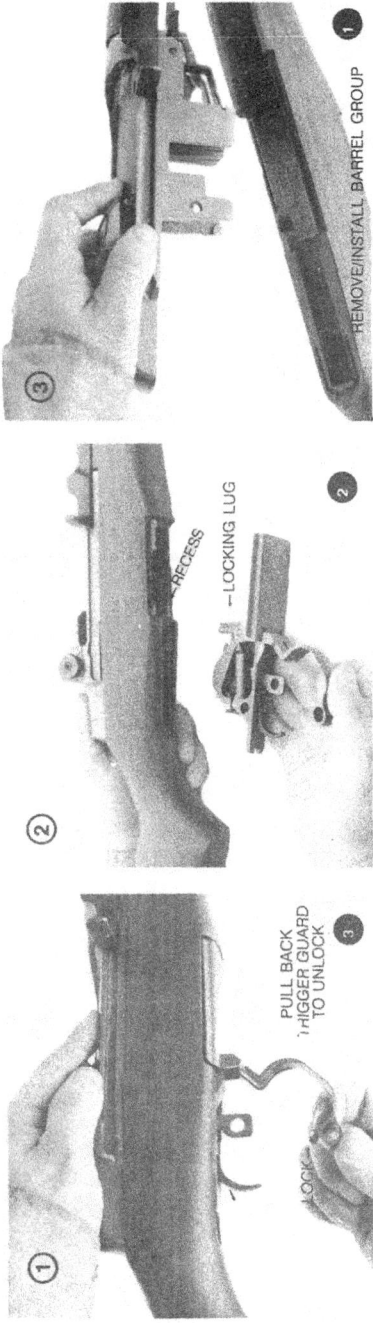

Figure 15. *Removal / installation of major groups and assemblies.*

REMOVE/INSTALL BARREL GROUP

RECESS

—LOCKING LUG

PULL BACK TRIGGER GUARD TO UNLOCK

LOCK

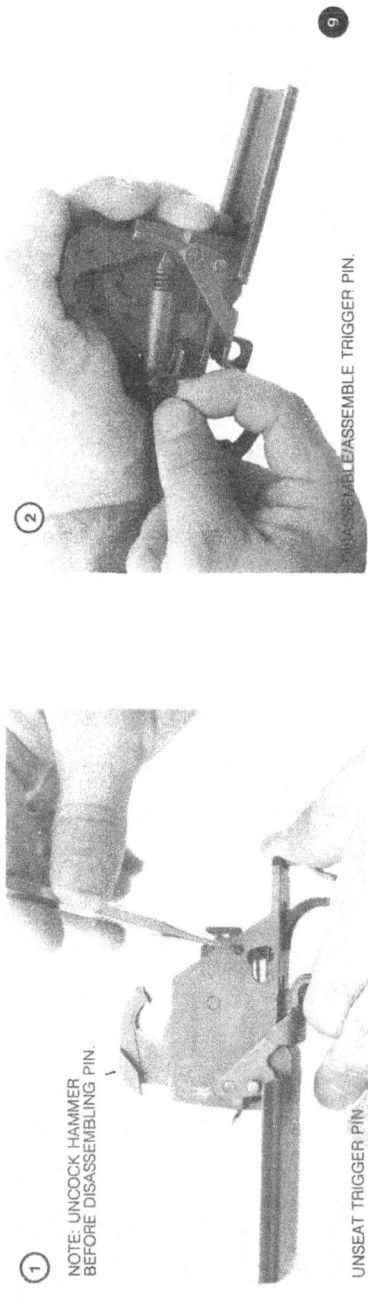

Figure 16. *Disassembly / assembly of trigger housing assembly (1 of 4).*

NOTE: UNCOCK HAMMER BEFORE DISASSEMBLING PIN.

UNSEAT TRIGGER PIN.

DISASSEMBLE/ASSEMBLE TRIGGER PIN.

27

PLUNGER

SPRING

HOUSING

DISASSEMBLE/ASSEMBLE HAMMER SPRING HOUSING,
HAMMER HELICAL COMPRESSION SPRING AND
HAMMER SPRING PLUNGER.

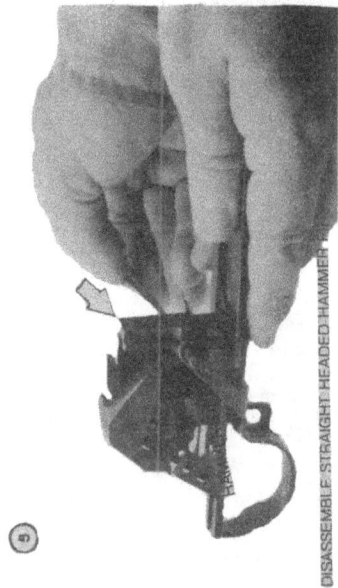

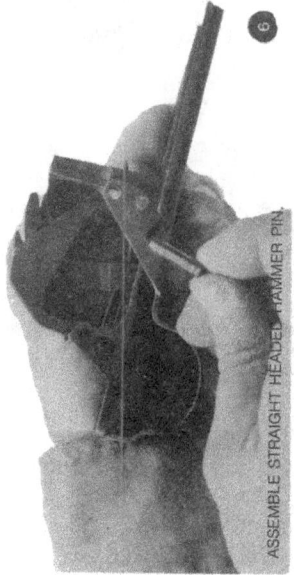

ASSEMBLE STRAIGHT HEADED HAMMER PIN.

TRIGGER

TRIGGER PIN

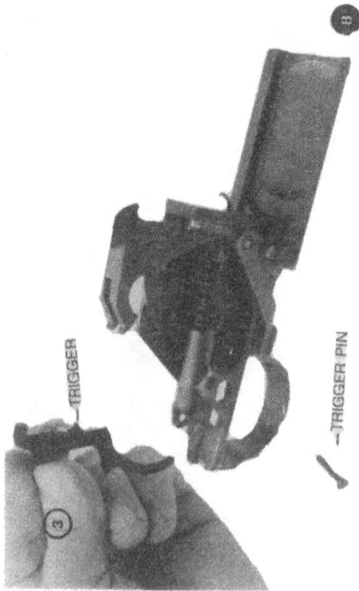

DISASSEMBLE/ASSEMBLE TRIGGER

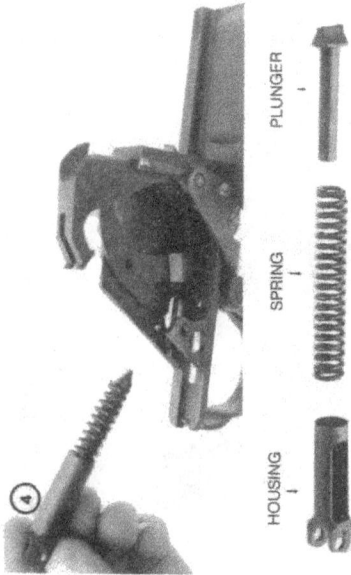

DISASSEMBLE STRAIGHT HEADED HAMMER

Figure 17. Disassembly / assembly of trigger housing assembly (2 of 4).

⑦ DISASSEMBLE/ASSEMBLE SAFETY.

④

⑥ DISASSEMBLE/ASSEMBLE HAMMER.

⑤

⑨ UNSEATING EJECTOR.

⑧ DISASSEMBLE/ASSEMBLE TRIGGER GUARD.

POSITION TRIGGER GUARD

ROTATE TRIGGER GUARD

③

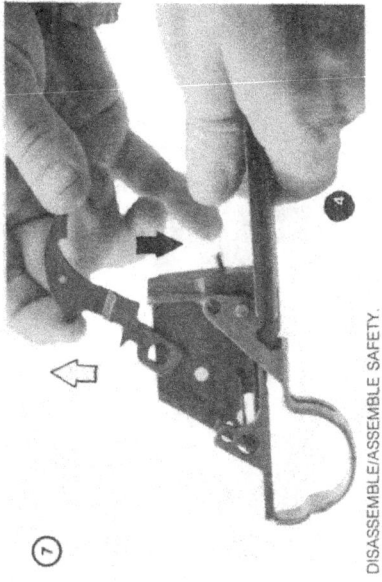

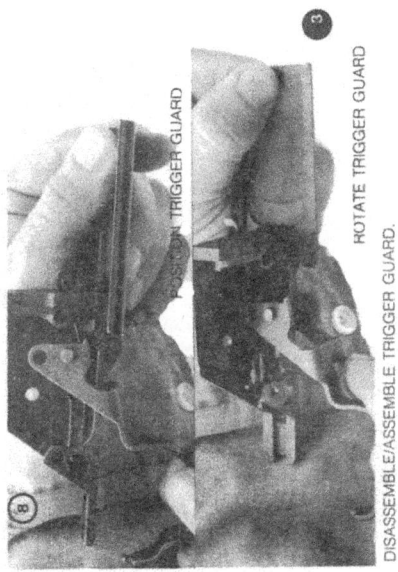

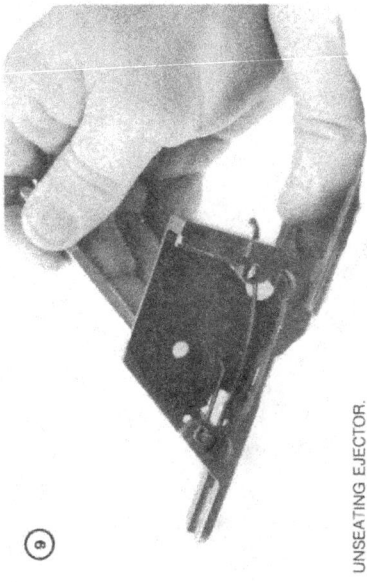

Figure 18. Disassembly / assembly of trigger housing assembly (3 of 4).

TRIGGER HOUSING

DISASSEMBLE/ASSEMBLE CLIP EJECTOR

SEAT CLIP EJECTOR ON STUD

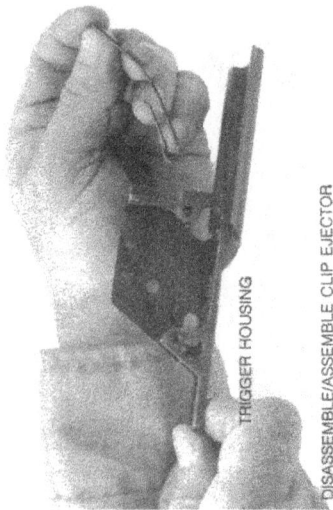

Figure 19. Disassembly / assembly of trigger housing assembly (4 of 4).

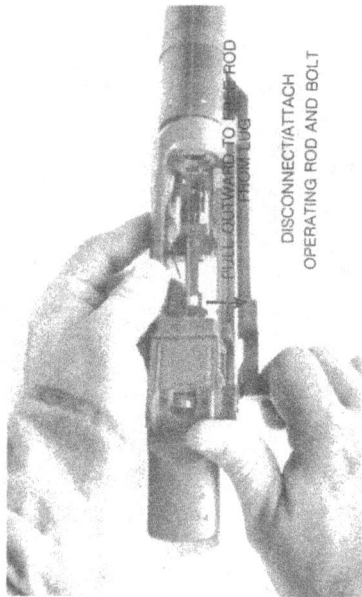

TO INSTALL,
PLACE TANG
ON TOP OF
RECEIVER BRIDGE.

OPERATING ROD GUIDE

PULL OUTWARD TO DETACH ROD
FROM LUG

DISCONNECT/ATTACH
OPERATING ROD AND BOLT

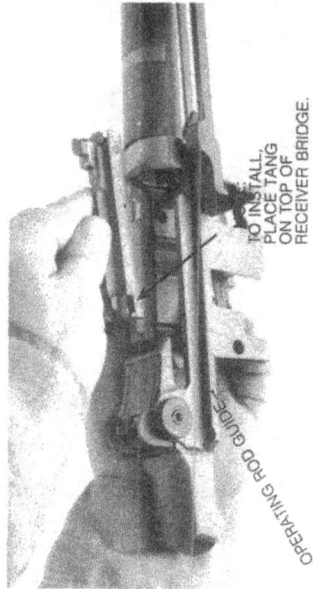

Figure 20. Removal / installation of bolt assembly.

DISASSEMBLY: PLACE TOOL IN BOLT AS SHOWN.
APPLY PRESSURE AND TURN CLOCKWISE TO
REMOVE EXTRACTOR.
NOTE: BE CAREFUL TO PREVENT SPRING AND
PLUNGER ASSEMBLIES FROM FLYING.

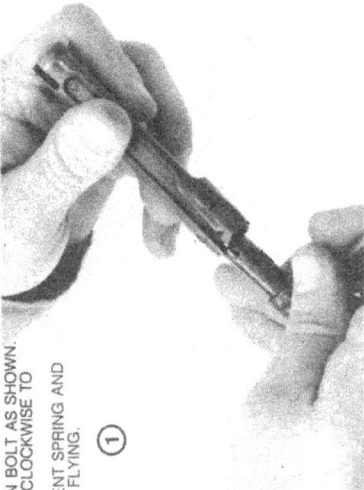

ASSEMBLY: HOLD BOLT IN HAND WITH THUMB ON
EXTRACTOR. PLACE TOOL SO GROOVE IS OVER
EJECTOR AND COMPRESS EJECTOR AND SPRING INTO
BOLT. PUSH EXTRACTOR IN PLACE WITH THUMB.

① DISASSEMBLE/ASSEMBLE EXTRACTOR WITH COMBINATION TOOL.

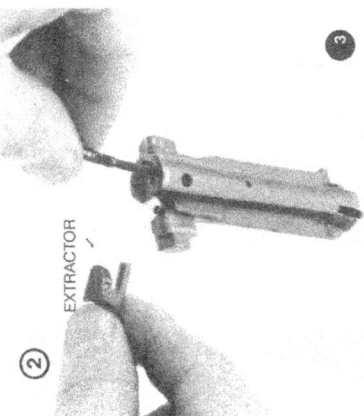

② DISASSEMBLE/ASSEMBLE EXTRACTOR AND CARTRIDGE EJECTOR.

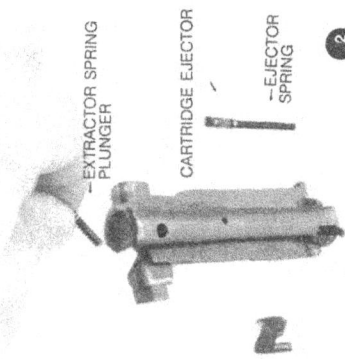

EXTRACTOR

③ DISASSEMBLE/ASSEMBLE EXTRACTOR SPRING PLUNGER.

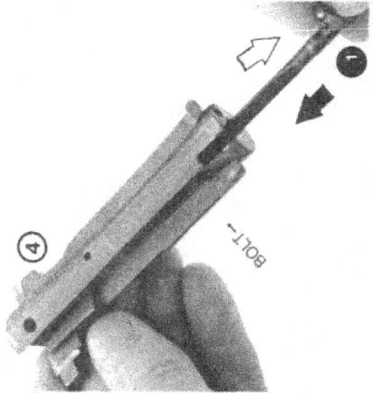

— EXTRACTOR SPRING
PLUNGER

CARTRIDGE EJECTOR

— EJECTOR
SPRING

④ DISASSEMBLE/ASSEMBLE FIRING PIN.

BOLT—

Figure 21. Disassembly / assembly of bolt assembly.

31

NOTE: USE COMBINATION TOOL TO LOOSEN
GAS CYLINDER LOCK SCREW.

DISASSEMBLE/ASSEMBLE GAS CYLINDER LOCK SCREW.

DISASSEMBLE/ASSEMBLE GAS CYLINDER LOCK.

DISASSEMBLE/ASSEMBLE FLASH HIDER T37 FROM
RIFLE M 1C, M 1D.

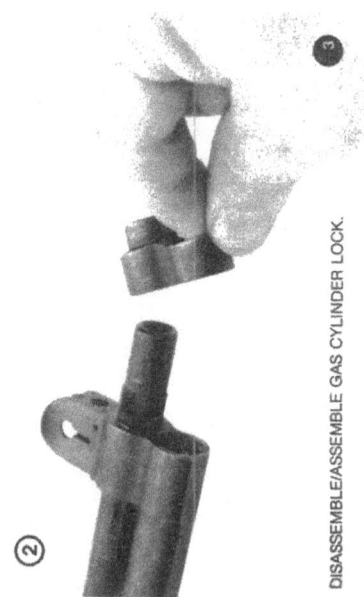

BARREL

LOOSEN GAS CYLINDER ASSEMBLY FROM BARREL.

Figure 22. Disassembly / assembly of gas cylinder group (1 of 2).

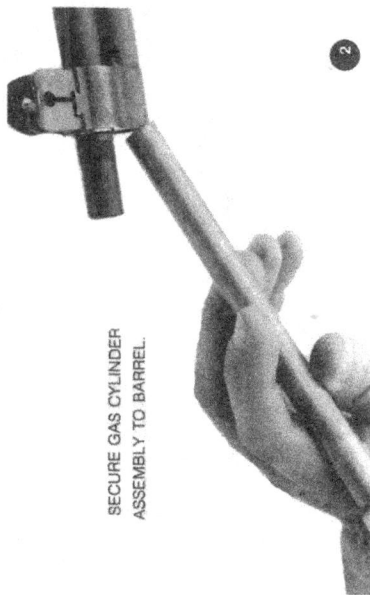

SECURE GAS CYLINDER
ASSEMBLY TO BARREL.

REMOVE/INSTALL GAS CYLINDER ASSEMBLY.

Figure 23. Disassembly / assembly of gas cylinder group (2 of 2).

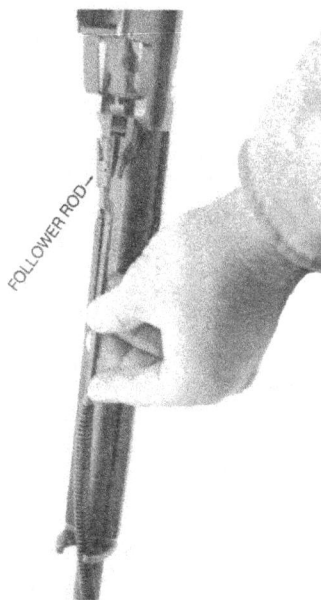

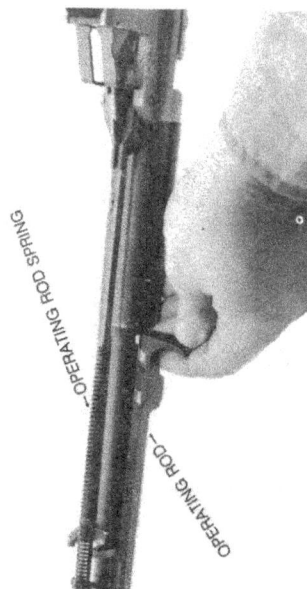

FOLLOWER ROD—

—OPERATING ROD SPRING

—OPERATING ROD

Figure 24. Removal / installation of operating rod assembly.

① DISASSEMBLE/ASSEMBLE SHOULDER HEADED PIN.

② FOLLOWER ASSEMBLY

DISASSEMBLE/ASSEMBLE FOLLOWER ARM.

③ DISASSEMBLE/ASSEMBLE BULLET GUIDE.

④ NOTE: ARM OF CATCH MUST GO IN UNDER LATCH

ARM—

REMOVE/INSTALL OPERATING ROD CATCH

⑤ REMOVE/INSTALL FOLLOWER ASSEMBLY.

⑥ NOTE: DISASSEMBLY OF FOLLOWER ASSEMBLY IS NECESSARY ONLY IF REPAIR OR REPLACEMENT OF PARTS IS REQUIRED

—SLOTTED END

FOLLOWER SLIDE—

DISASSEMBLE FOLLOWER SLIDE FROM FOLLOWER.

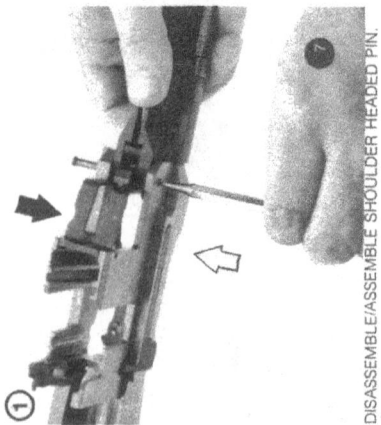

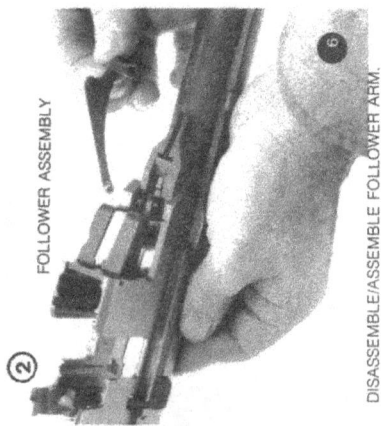

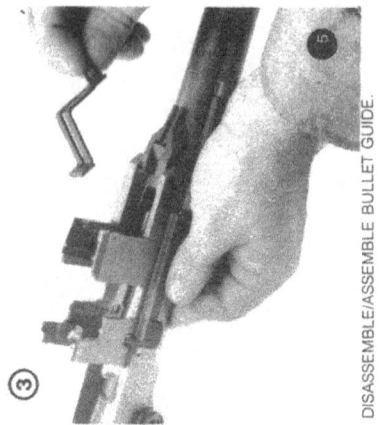

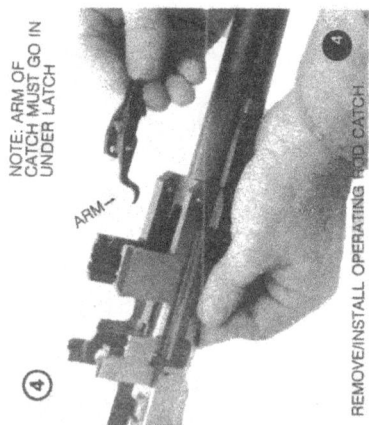

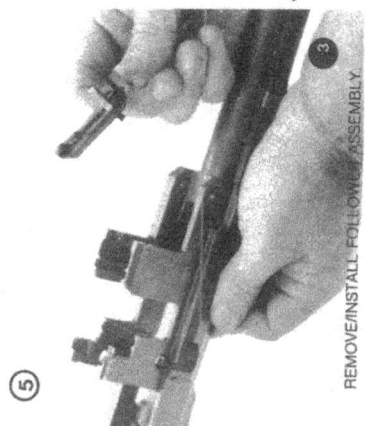

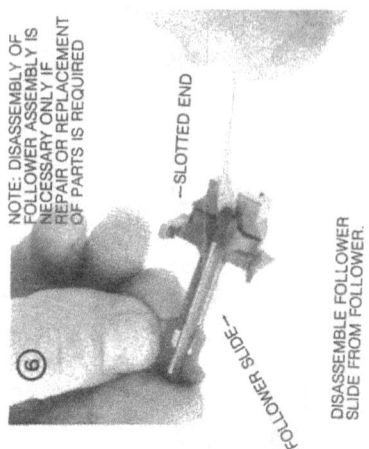

Figure 25. Disassembly / assembly of magazine follower group (1 of 2).

FOLLOWER SLIDE

FOLLOWER

FOLLOWER SLIDE AND FOLLOWER DISASSEMBLED.

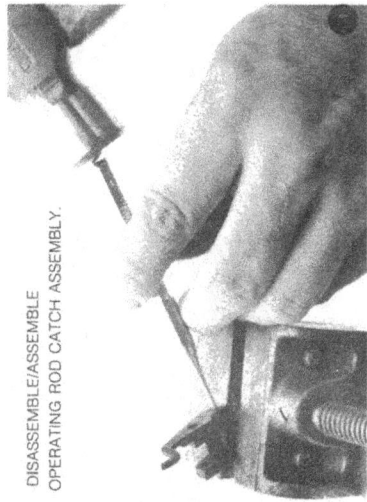

ACCELERATOR

OPERATING ROD CATCH

ARM

PIN

OPERATING ROD CATCH ASSEMBLY DISASSEMBLED

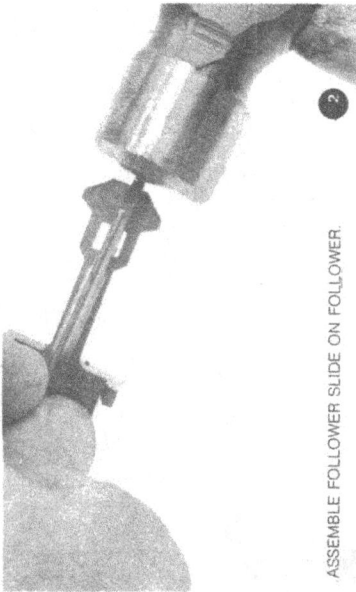

② ASSEMBLE FOLLOWER SLIDE ON FOLLOWER.

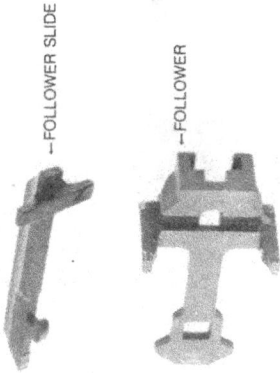

⑦ DISASSEMBLE/ASSEMBLE OPERATING ROD CATCH ASSEMBLY.

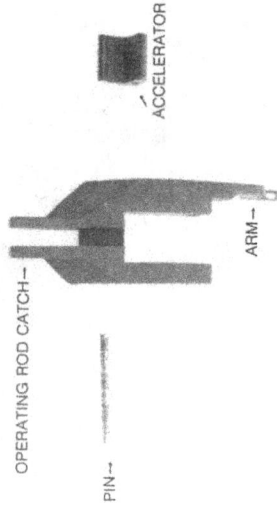

Figure 26. Disassembly / assembly of magazine follower group (2 of 2).

35

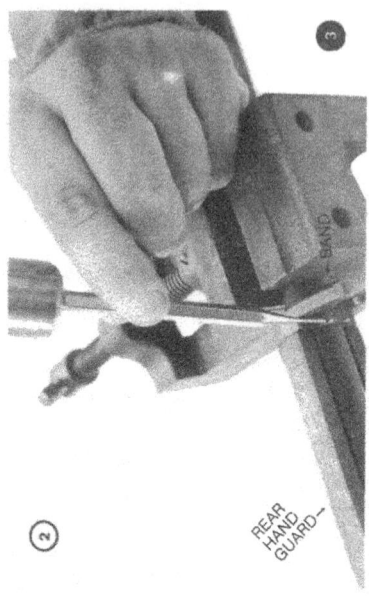

REAR
HAND–
GUARD

DISASSEMBLE/ASSEMBLE SPRING PIN

DISASSEMBLE/ASSEMBLE BAND

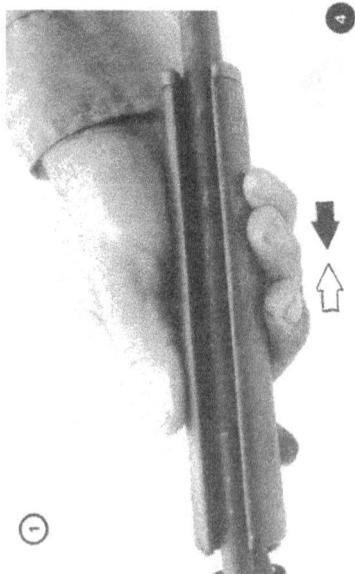

DISASSEMBLE/ASSEMBLE FRONT HAND GUARD

LOOSEN/TIGHTEN BAND

Figure 27. Disassembly / assembly of handguard group (1 of 2).

DISASSEMBLE/ASSEMBLE REAR HAND GUARD

BARREL

SPRING PIN

BAND

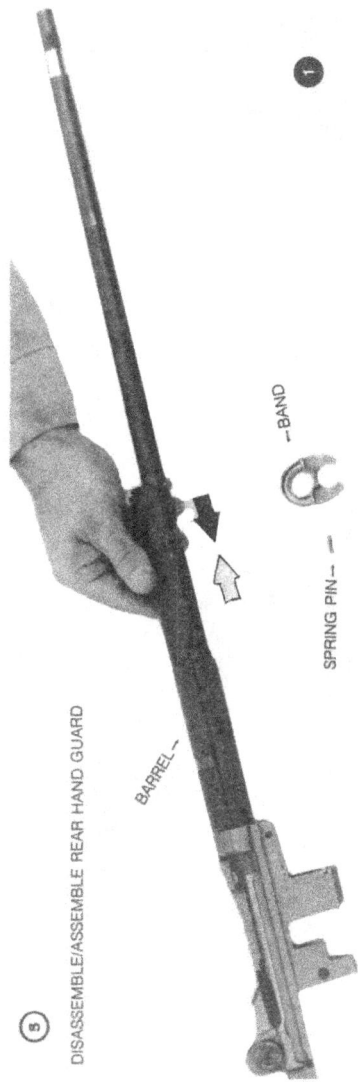

Figure 28. Disassembly / assembly of handguard group (2 of 2).

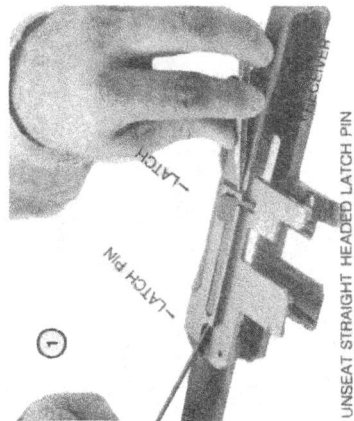

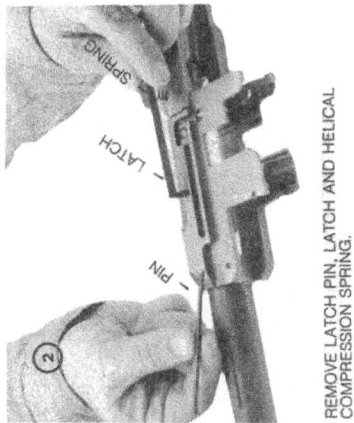

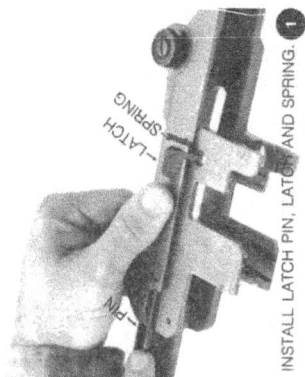

UNSEAT STRAIGHT HEADED LATCH PIN

LATCH PIN

LATCH

REMOVE LATCH PIN, LATCH AND HELICAL COMPRESSION SPRING.

LATCH

SPRING

PIN

INSTALL LATCH PIN, LATCH AND SPRING.

LATCH SPRING

PIN

Figure 29. Disassembly / assembly of latch group.

37

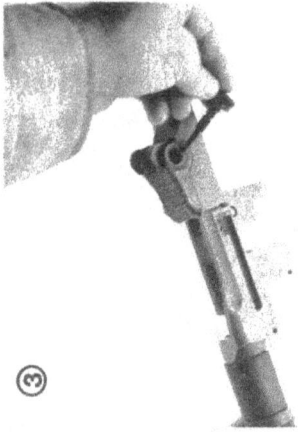

③ DISASSEMBLE/ASSEMBLE ELEVATING PINION.

BASE

⑥ DISASSEMBLE/ASSEMBLE REAR SIGHT COVER AND BASE.

—KNOB

② DISASSEMBLE/ASSEMBLE WINDAGE KNOB.

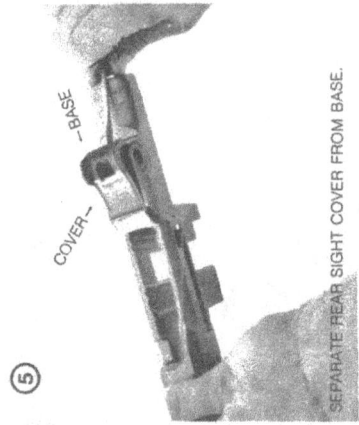

COVER— —BASE

⑤ SEPARATE REAR SIGHT COVER FROM BASE.

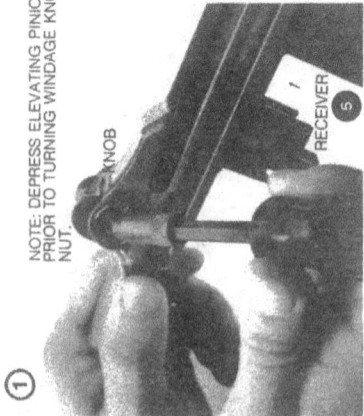

NOTE: DEPRESS ELEVATING PINION (LEFT SIDE) PRIOR TO TURNING WINDAGE KNOB LOCKING NUT.

KNOB

RECEIVER

① LOOSEN/TIGHTEN NUT.

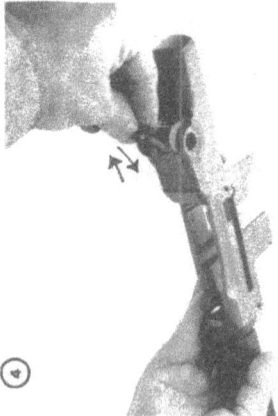

④ DISASSEMBLE/ASSEMBLE APERTURE.

Figure 30. Disassembly / assembly of rear sight group.

38

Chapter 4
MAINTENANCE OF MATERIEL USED IN CONJUNCTION WITH MAJOR ITEM

4-1. General
Refer to table 4-1.

Table 4-1. Maintenance of Equipment

Item	Maintenance function	
	Operator's maintenance	Organizational maintenance
Grenade Launcher, M7A3	Remove/install (fig. 31). Clean/lubricate.	Repair/replace. Refer to TM 9-1005-234-14P for authorized parts.
Grenade Launcher Sight, M15	Remove/install (fig. 31). Clean/lubricate.	Repair/replace. Refer to TM 9-1005-234-14P for authorized parts.
Bayonet-Knife, M5 and M5A1	Remove/install (fig 31). Clean/lubricate.	Repair/replace. Refer to TM 9-1005-237-15P for authorized parts.
Scabbard, M8A1	Clean.	Repair/replace. Refer to TM 9-1005-237-15P for authorized parts.
Winter trigger kit	Clean/lubricate.	Repair/replace. Adjust trigger bar. Refer to appendix B for authorized parts.

INSTALL/REMOVE BAYONET
INSTALL
COMPRESS TO REMOVE— BAYONET
REMOVE—

INSTALL/REMOVE GRENADE LAUNCHER
INSTALL
REMOVE
INSTALL
REMOVE→

POSITIONING SIGHT ON MOUNTING PLATE.

ALIGNING CLICK SPRING TIPS TO MOUNTING PLATE NOTCHES.

SIGHT, M15 INSTALLED

ELEVATING SCREW—

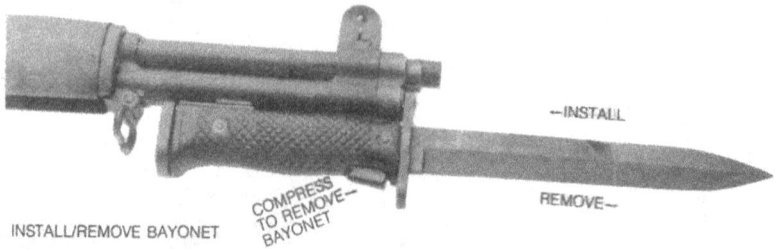

Figure 31. Materiel used in conjunction with major items.

Chapter 5
AMMUNITION

Ammunition is loaded into 8 round clips, which are inserted, en bloc, into the magazine. (see fig. 32).

5-1. Types
Refer to SC 1305/30 IL for identification of various types of ammunition.

5-2. Care, handling, Preservation, and Destruction
Refer to TM 9-1300-206.

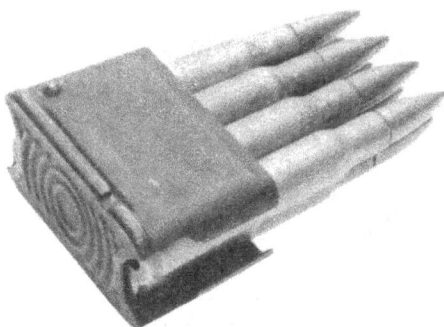

Figure 32. Cartridges and clip.

Chapter 6
DESTRUCTION OF MATERIEL TO PREVENT ENEMY USE

6-1. General
a. Destruction of the rifle when subject to capture or abandonment in the combat zone, will be undertaken only when in the judgment of the commander concerned such action is necessary. If destruction is resorted to, the equipment must be so badly damaged that it cannot be restored to a usable condition in the combat zone either by repair or cannibalization. The reporting of the destruction of equipment is to be through regular channels.

b. Priorities for destruction of repair parts are:
(1) Firing pin
(2) Extractor
(3) Ejector
(4) Hammer spring
(5) Tigger
(6) Safety

Appendix A
REFERENCES

A-1. Publication Indexes
Consult the following publication indexes frequently for the latest changes or revisions of references and for new publications relating to material covered in this manual.

A-2. Forms
DA Form 2028, Recommended Changes to DA Publications
DA Form 2407, Maintenance Request
DD Form 6, Report of Damaged or Improper Shipment
DA Form 9-79, Parts Requisition

A-3. Other Publications
The following explanatory publications pertain to this material.

a. General

b. Ammunition.

c. Inspection and Maintenance.

d. Training.

Appendix B
ORGANIZATIONAL MAINTENANCE REPAIR PARTS
AND SPECIAL TOOLS LIST

Section I. INTRODUCTION

B-1. Scope

This appendix lists basic issue items, repair parts, and special tools required for the performance of organizational maintenance of the Rifles M1, M1C (Sniper's) and M1D (Sniper's).

B-2. General

The Basic Issue Items, Repair Parts, and Special Tools List is divided into the following sections:

a. Basic Issue Items List-Section II. A list of items which accompany the rifle and are required by the operator/crew for installation, operation, or maintenance.

b. Maintenance and Operating Supplies-Section III. A listing of maintenance and operating supplies required for initial operation.

c. Prescribed Load Allowance (PLA)-Section IV. A composite listing of repair parts, special tools, test and support equipment having quantitative allowances for initial stockage at the organizational level.

d. Repair Parts-Section V. A list of repair parts authorized for the performance of maintenance at the organizational level in figure and item number sequence.

e. Special Tools, Test and Support Equipment-Section VI. A list of special tools, test and support equipment authorized for the performance of maintenance at the organizational level.

f. Federal Stock Number and Reference Number Index-Section VII. A list of Federal stock numbers in ascending numerical sequence, followed by a list of reference numbers, appearing in all the listings, in ascending alpha-numeric sequence, cross-referenced to the illustration figure number and item number.

B-3. Explanation of Columns

The following provides an explanation of columns in the tabular lists in Sections II through VI.

a. Source, Maintenance, and Recoverability Codes (SMR).

(1) Source Code. Indicates the selection status and source for the listed item. Source codes used are:

Code	Explanation
P	Repair parts which are stocked in or supplied from the GSA/DSA, or Army supply system, and authorized for use at indicated maintenance categories.
P2	Repair parts which are procured and stocked for insurance purposes because the combat or military essentiality of the end item dictates that a minimum quantity be available in the supply system.
M	Repair parts which are not procured or stocked but are to be manufactured in indicated maintenance levels.
A	Assemblies which are not procured or stocked as such but are made up of two or more units. Such component units carry individual FSN's and descriptions, are procured and stocked separately and can be assembled to form the required assembly at indicated maintenance categories.
X	Parts and assemblies which are not procured or stocked and the mortality of which is normally below that of the applicable end item or component. The failure of such part or assembly should result in retirement of the end item from the supply system.
X1	Repair parts which are not procured or stocked. The requirement for such items will be filled by use of the next higher assembly or component.

| X2 | .... | Repair parts which are not stocked. The indicated maintenance category requiring such repair parts will attempt to obtain through cannibalization; if not obtainable through cannibalization, such repair parts will be requisitioned with supporting justification through normal supply channels. |
| G | .... | Major assemblies that are procured with PEMA funds for issue initial only to be used as exchange assemblies at DSU and GSU level. These assemblies will not be stocked above DSU and GSU level or returned to Depot supply level. |

(2) Maintenance Code. Indicates the lowest category of maintenance authorized to install the item. The maintenance level codes are:

Code	Explanation
C	Crew or operator
O	Organizational
F	Direct Support
H	General Support
D	Depot

(3) Recoverability Code. Indicates whether unserviceable items should be returned for recovery or salvage. Items not coded are expendable. The recoverability code is.

Code	Explanation
R	Repair parts and assemblies which are economically repairable at DSU and GSU activities and are normally furnished by supply on an exchange basis.
S	Repair parts and assemblies which are economically repairable at DSU and GSU activities and normally are furnished by supply on an exchange basis. When items are determined by a GSU to be uneconomically repairable, they will be evacuated to a depot for evaluation and analysis before final disposition.
T	High dollar value recoverable repair parts which are subject to special handling and are issued on an exchange basis. Such repair parts are normally repaired or overhauled at depot maintenance activities.
U	Repair parts specifically selected for salvage by reclamation units because of precious metal content, critical materials, high dollar value reusable casings, or castings.

No Code
indicated-Part will be considered expendable.

b. Federal Stock Number. Indicates the Federal stock number assigned to the item and will be used for requisitioning purposes.

c. Description. Indicates the Federal item name and any additional description of the item required. The abbreviation "w/e" when used as a part of the nomenclature, indicates the Federal stock number includes all armament, equipment, accessories, and repair parts issued with the item. A part number or other reference number is followed by the applicable five-digit Federal supply code for manufacturers in parentheses.

d. Unit of Measure (U/M). A 2 character alphabetic abbreviation indicating the amount or quantity of the item upon which the allowances are based, e.g., ft, ea, pr, etc.

e. Quantity Incorporated in Unit. Indicates the quantity of the item used in the functional group or assembly. A"V" appearing in this column in lieu of a quantity indicates that a definite quantity cannot be indicated (e.g., shims, spacers, etc.).

f. Quantity Furnished with Equipment. Indicates the quantity of an item furnished with the equipment (BIIL only).

g. Component Application. Identifies the component application of each maintenance or operating supply item (M&O supplies only).

k. 15-Day Organizational Maintenance Allowances.

(1) The allowance columns are divided into four subcolumns. Indicated in each sub-column opposite the first appearance of each item is the total quantity of items authorized for the number of equipments supported. Subsequent appearances of the same item will have the letters "REF" in the allowance columns. Items authorized for use as required but not for initial stockage are identified with an asterisk in the allowance column.

(2) The quantitative allowances for organizational level of maintenance represents one initial prescribed load for a 15-day period for the number of equipments supported. Units and organizations authorized additional prescribed loads will multiply the number of prescribed loads authorized by the quantity of repair parts reflected in the appropriate density column to obtain the total quantity of repair parts authorized.

(3) Organizational units providing maintenance for more than 100 of these equipments shall determine the total quantity of parts required by converting the equipment quantity to a decimal factor by placing a decimal point before the next to last digit of the number to indicate hundredths, and multiplying the decimal factor by the parts quantity authorized in the 51-100 allowance column. Example, authorized allowance for 51-100 equipments is 12; for 140 equipments multiply 12 by 1.40 or 16.80 rounded off to 17 parts required.

(4) Subsequent changes to allowances will be limited as follows: No change in the range of items as follows: No change in the range of items is authorized. If additional items are considered necessary, recommendations should be forwarded to Commanding General, Headquarters, U.S. Army Weapons Command, ATTN: AMSWE-SMM-SA, Rock Island, Illinois 61201, for exception or revision to the allowance list. Revisions to the range of items authorized will be made by the U.S. Army Weapons Command based upon engineering experience, demand data, or TAERS information.

l. Illustration.

(1) **Figure Number.** Indicates the figure number of the illustration in which the item is shown.

(2) **Item Number.** Indicates the callout number used to reference the item in the illustration.

Note. Items called-out on illustration, but not listed, are for disassembly purposes only.

B-4. Special Information

Identification of the usable on codes of this publication are:

Code	Used on
No Code ...	M1, M1C, and M1D
A	M1
B	M1C
C	M1D
D	M1 and M1C
E	M1 and M1D
F	M1C and M1D

B-5. How to Locate Repair Parts

a. When Federal stock number or reference number is unknown.:

(1) First. Using the table of contents determine the functional group or assembly, within which the repair part belongs. This is necessary since illustrations are prepared for functional groups and assemblies, and listings are divided into the same groups.

(2) Second. Find the illustration covering the functional group or assembly to which the repair part belongs.

(3) Third. Identify the repair part on the illustration and note the illustration figure and item number of the repair part.

(4) Fourth. Using the Repair Parts Listing, find the functional group or assembly to which the repair part belongs and locate the illustration figure and item number noted in the illustration.

b. When Federal stock number or reference number is known:

(1) First. Using the Index of Federal Stock Numbers and Reference Numbers find the pertinent Federal stock number or reference number. This index is in ascending FSN sequence followed by a list of reference numbers in alpha-numeric sequence, cross-referenced to the illustration figure number and item number.

(2) Second. Using the Repair Part Listing, find the functional group or assembly of the repair part and the illustration figure number and item number referenced in the Index of Federal Stock Numbers and Reference Numbers.

B-6. Federal Supply Codes for Manufacturers

Code	Manufacturer
19200	Frankford Arsenal
19204	Rock Island Arsenal
19205	Springfield Armory

Section II. BASIC ISSUE ITEMS LIST

(2)	(3)	(4)	(5)	(6)	(7) Illustration	
Federal Stock No.	Description Reference Number & Mfr. Code	Unit of issue	Qty. inc. in unit	Qty. furn. with equip.	(a) Fig. No.	(b) Item No.
	RIFLES, CALIBER .30, M1, M1C (SNIPER'S) AND M1D (SNIPER'S) **REPAIR PARTS:** NONE AUTHORIZED					
	TOOLS AND EQUIPMENT					
1005-556-4174	BRUSH, CLEANING, SMALL ARMS: BORE 5564174 (19205)	EA	--	1	33	1
1005-691-1381	BRUSH, CLEANING, SMALL ARMS: CHAMBER 7790582 (19205)	EA	--	1	33	2
1005-791-3377	CASE LUBRICANT: 7790995 (19205)	EA	--	1	33	7
1005-650-4510	CASE, SMALL ARMS CLEANING ROD: 7267754 (19204)	EA	--	1	33	6
1005-793-6761	HANDLE ASSEMBLY: CLEANING ROD: 7266115 (19204)	EA	--	1	33	3
1005-726-6109	ROD SECTION, CLEANING, SMALL ARMS: 7266109 (19205)	EA	--	4	33	4
1005-654-4058	SLING, SMALL ARMS: 6544058 (19205)	EA	--	1	34	
1005-726-6110	SWAB HOLDER SECTION, SMALL ARMS CLEANING ROD: 7266110 (19204)	EA	--	1	33	5

Section III. MAINTENANCE AND OPERATING SUPPLIES

(1) Component application	(2) Federal stock number	(3) Description
CALIBER .30 RIFLES M1, M1C (Sniper's) and M1D (Sniper's)	1005-288-3565	SWAB, SMALL ARMS CLEANING: COTTON 2½ sq (1,000 IN PKG) 5019316 (19204)

Section IV. PRESCRIBED LOAD ALLOWANCE

(1) Federal stock No.	(2) Description Usable on code	(3) Qty.inc. in unit pack	(4) 15-day organizational maint. allowance			
			(a) 1-5	(b) 6-20	(c) 21-50	(d) 51-100
	REPAIR PARTS:					
1005-313-9441	SCREW	~~	~~	2	2	2
1005-501-3667	PIN, SHOULDER, HEADED	~~	~~	~~	2	2
1005-554-6015	SAFETY, SMALL ARMS	~~	~~	~~	2	2
1005-554-6018	EJECTOR	~~	~~	~~	~~	2
1005-554-6024	GUARD, HAND GUN D	~~	~~	~~	2	2
1005-554-6026	TRIGGER	~~	~~	~~	~~	2
1005-556-4245	GUARD, HAND GUN	~~	~~	2	2	2
1005-600-8616	EJECTOR, CARTRIDGE	~~	~~	~~	2	2
1005-600-6818	PLUNGER, EXTRACTOR SPRING	~~	~~	2	2	2
1005-600-8868	APERTURE, SIGHT	~~	~~	2	2	2
1005-600-8879	PIN, FIRING	~~	~~	~~	2	2
1005-600-8885	SPRING, HELICAL, COMPRESSION	~~	~~	~~	2	2
1005-600-8887	SPRING, HELICAL, COMPRESSION	~~	~~	2	2	2
1005-600-8891	SWIVEL, STACKING	~~	~~	2	2	2
1005-614-7568	SPRING, HELICAL, COMPRESSION	~~	~~	~~	2	2
1005-731-2556	GUARD, HAND, GUN C	~~	~~	2	2	2
1005-731-2737	KNOB	~~	~~	2	2	2
1005-819-4501	PIN, TRIGGER	~~	~~	2	2	2
1005-953-9504	EXTRACTOR, CARTRIDGE	~~	~~	2	2	2
1005-999-3400	PINION	~~	~~	2	2	3
5305-501-3678	SCREW, MACHINE	~~	~~	~~	2	2
5315-501-3668	PIN, STRAIGHT, HEADED	~~	~~	2	2	3
	TOOLS AND EQUIPMENT:					
1005-288-3565	SWAB, SMALL ARMS CLEANING	~~	~~	2	2	2
1005-556-4174	BRUSH, CLEANING, SMALL ARMS	~~	~~	2	3	6
1005-650-4510	CASE, SMALL ARMS CLEANING ROD	~~	~~	2	2	2
1005-654-4058	SLING, SMALL ARMS	~~	~~	2	2	3
1005-691-1381	BRUSH CLEANING, SMALL ARMS	~~	~~	2	2	3
1005-694-1662	BUFFER, CLEANING ROD	~~	~~	2	2	3
1005-726-6109	ROD SECTION, CLEANING, SMALL ARMS	~~	~~	2	2	2
1005-726-6110	SWAB HOLDER SECTION, SMALL ARMS CLEANING ROD	~~	~~	2	2	3
1005-791-3377	CASE, LUBRICANT	~~	~~	2	2	2
1005-793-6761	HANDLE ASSEMBLY	~~	~~	2	2	2
1240-763-1596	CASE, TELESCOPE	~~	~~	~~	2	2

Section V. REPAIR PARTS LIST

(1) Source maint. recov. code			(2) Federal stock No.	(3) Description — Reference Number & Mfr Code / Usable on Code	(4) Unit of means / unit	(5) Qty inc in unit	(6) 15 day organizational maintenance alw				(7) Illustration	
(a)	(b)	(c)					(a) 1-5	(b) 6-20	(c) 21-50	(d) 51-100	(a) Figure No.	(b) Item No.
				REPAIR PARTS FOR: RIFLES, CALIBER .30, M1, M1C (SNIPER'S) AND M1D (SNIPER'S)								
				TRIGGER HOUSING ASSEMBLY								
P	O	~	1005-819-4501	PIN, TRIGGER: 7791367 (19205)	EA	1	*	1	2	3	35	1
P	O	~	1005-554-6026	TRIGGER: 5546026 (19205)	EA	1	*	*	*	2	35	2
P	O	~	1005-600-8887	SPRING, HELICAL, COMPRESSION: 20 COILS, HAMMER 6008887 (19205)	EA	1	*	2	2	2	35	4
P	O	~	5315-501-3668	PIN, STRAIGHT, HEADED: S, FL-FIL-HD, 0.187 MAX DIA SHANK, ¾ NOM LG UNDER 5013668 (19204)	EA	1	*	2	2	3	B-2	6
P	O	~	1005-554-6015	SAFETY SMALL ARMS:	EA	1	*	*	2	2	35	8/8W
P	O	~	1005-554-6018	EJECTOR: CLIP 5546018 (19205)	EA	1	*	*	*	2	35	10
				BARREL AND RECEIVER GROUP:								
P	O	~	1005-614-7568	SPRING, HELICAL, COMPRESSION: 200 TOTAL COILS, OPERATING ROD 6147568 (19205)	EA	1	*	*	2	2	36	2
A	~	~	----------	BOLT ASSEMBLY 5546023	~	~	~	~	~	~	36	4
P	O	~	1005-953-9504	EXTRACTOR, CARTRIDGE:	EA	1	*	2	2	2	36	5
P	O	~	1005-600-8616	EJECTOR, CARTRIDGE:	EA	1	*	*	2	2	36	6
P	O	~	1005-600-8618	PLUNGER, EXTRACTOR SPRING: 6008618 (19205)	EA	1	*	2	2	2	36	7
P	O	~	1005-600-8879	PIN, FIRING: BOLT ASSY 6008879 (19205)	EA	1	*	*	2	2	36	8
P	O	~	1005-501-3667	PIN, SHOULDER, HEADED: FOLLOWER ARM 5013667 (19204)	EA	1	*	*	2	2	36	10
P	O	~	1005-313-9441	SCREW: GAS CYLINDER LOCK 7267797 (19205)	EA	1	*	2	2	2	36	15

			Stock Number	Description	U/M	Qty				Fig	Item
P	O	~	5305-501-3678	SCREW, MACHINE: S, FIL-HD, NO. 10(0.190) -32NF-2 X 0.700 MAX LG (STACKING SWIVEL) 5013678 (19204)	EA	1	*	2	2	36	18
P	O	~	1005-600-8891	SWIVEL, STACKING: GAS CYLINDER 6008891 (19205)	EA	1	2	2	2	36	19
P	O	R	1005-556-4245	GUARD, HAND, GUN: FRONT D	EA	1	*	2	2	36	20
P	O	R	1005-554-6024	GUARD, HAND, GUN: REAR	EA	1	2	2	2	36	23
P	O	R	1005-731-2556	GUARD, HAND, GUN: REAR 7312556 (19205)	EA	1	*	2	2	36	24
P	O	~	1005-600-885	SPRING, HELICAL, COMPRESSION: 6-7/8 COILS (LATCH) 6008885 (19205)	EA	1	*	2	2	36	27
P	O	~	1005-731-2737	KNOB: WINDAGE	EA	1	2	2	2	36	28
P	O	~	1005-999-3400	PINION: ELEVATING 1010364 (19204)	EA	1	2	2	3	36	29
P	O	~	1005-600-8868	APERTURE, SIGHT: 6008868 (19204)	EA	1	2	2	2	36	30
P	O	~	1240-766-7454	TELESCOPE, M84 EYESHIELD: 7667454 (19200)	~	~	~	~	~	37	1
P	O	~		KIT, WINTER TRIGGER	EA	1	*	*	*	37	2
P	O	~	1005-775-0364	TRIGGER ASSEMBLY, WINTER: M5 7790808 (19205)	EA	1	*	*	*	38	1
P	O	~	5305-990-6435	SCREW, TAPPING, THREAD FORMING: 7791415 (19205)	EA	2	*	*	*	38	2
P	O	~	1005-010-5022	WASHER, HINGE RETAINING: TRIGGER ASSEMBLY 7791237 (19205)	EA	1	*	*	*	38	3
P	O	~	1005-778-0580	SAFETY, WINTER: 7790903 (19205)	EA	1	*	*	*	38	5

MATERIAL REQUIRED FOR COLD WEATHER CLIMATES

THE FOLLOWING ITEM IS ISSUED OR REQUISITIONED ONLY BY SPECIAL AUTHORIZATION OF AREA COMMANDER.

			Stock Number	Description	U/M	Qty
~	R	~	1005-777-1369	KIT, WINTER TRIGGER: FOR ARCTIC HANDWEAR 5910520 (19204)	EA	1

NOTE. INSTALLATION WILL BE PERFORMED BY DIRECT SUPPORT MAINTENANCE PERSONNEL.

Section VI. SPECIAL TOOLS, TEST AND SUPPORT EQUIPMENT

(1) Source maint. recov. code			(2) Federal stock No.	(3) Description / Reference Number & Mfr Code / Usable on Code	(4) Unit of meas	(5) Qty inc in unit	(6) 15 day organizational maintenance alw				(7) Illustration	
(a)	(b)	(c)					(a) 1-5	(b) 6-20	(c) 21-50	(d) 51-100	(a) Figure No.	(b) Item No.
				TOOLS AND EQUIPMENT								
				TOOLS AND EQUIPMENT AUTHORIZED FOR UNIT REPLACEMENT								
	C		1005-228-3536	SWAB, SMALL ARMS CLEANING: COTTON, 2½ SQ (1,000 IN PG) 5019916 (19204)	PG		*			2		
	C		1005-556-4174	BRUSH, CLEANING, SMALL ARMS: BORE 5564174 (19205)	EA		*	2	3	6	33	1
	C		1005-650-4510	CASE, SMALL ARMS CLEANING ROD: 7267754 (19205)	EA		*	2	2	2	33	6
	C		1005-654-4058	SLING, SMALL ARMS:	EA		*	2	2	3	34	2
	C		1005-691-1381	BRUSH, CLEANING, SMALL ARMS: CHAMBER 779583 (19205)	EA		*	2	2	3	33	2
	C		1005-726-6109	ROD SECTION, CLEANING, SMALL ARMS: 7266109 (19205)	EA		*	2	2	2	33	4
	C		1005-726-6110	SWAB HOLDER SECTION, SMALL ARMS CLEANING ROD: 7266110 (19204)	EA		*	2	2	3	33	5
	C		1005-791-3377	CASE, LUBRICANT: 790995 (19205)	EA		*	2	2	2	33	7
	C		1005-793-6761	HANDLE ASSEMBLY: CLEANING ROD 7266115 (19204)	EA		*	2	2	2	33	3
				ORGANIZATIONAL MAINTENANCE TOOLS AND EQUIPMENT (FOR ARMORERS USE) THE 15-DAY LEVEL IS NOT APPLICABLE.								
	O		1005-722-8907	ENVELOPE: FABRIC, 2-BUTTON, 3H X 4-7/8W 2228907 (19205)	EA		1				40	1
	O		4933-628-9700	REFLECTOR, GUN BARREL. 7790138 (19205)	EA		1				40	2
	O		4933-652-9950	EXTRACTOR, RUPTURED CARTRIDGE CASE: 7790352 (19205)	EA		1				40	3

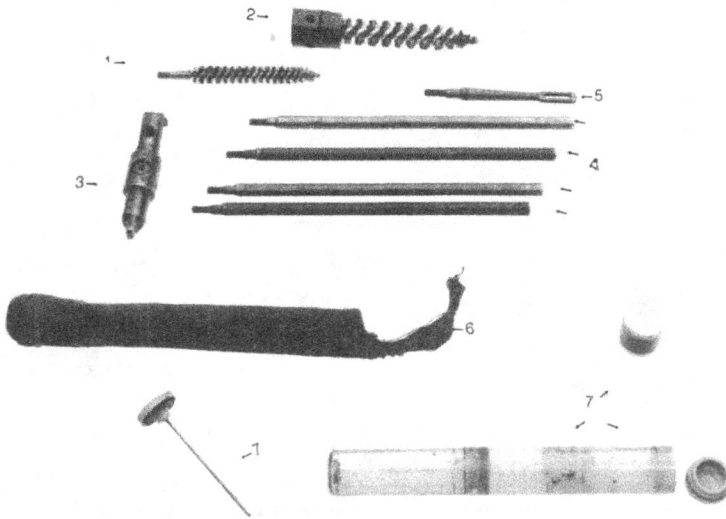

Figure 33. Basic cleaning tools.

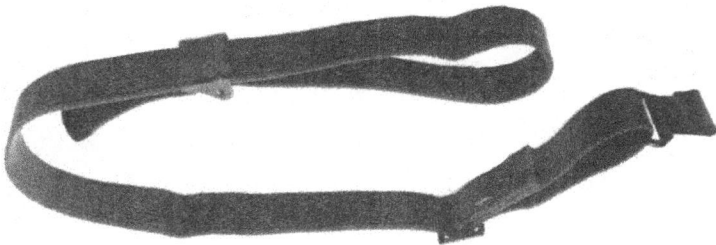

Figure 34. Web sling.

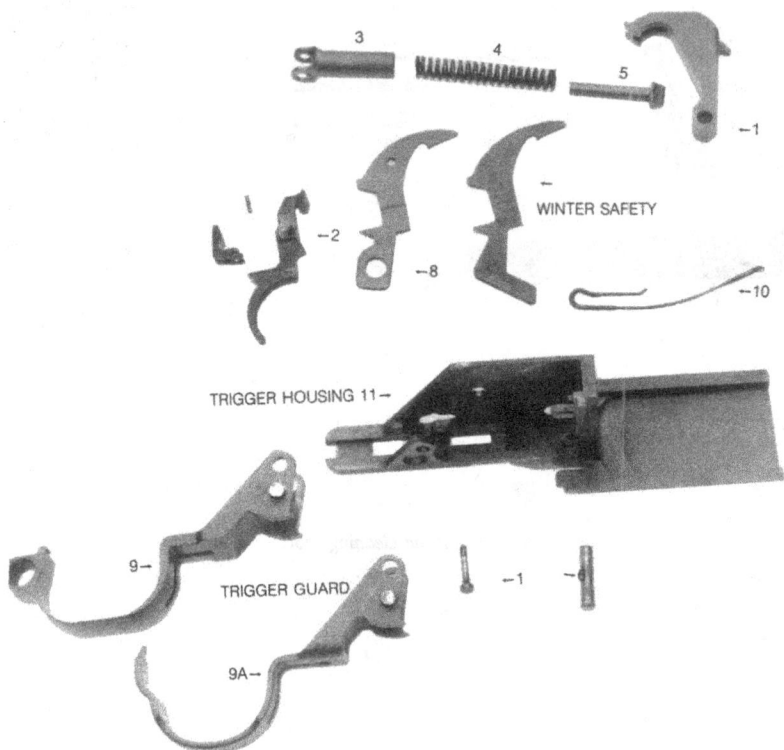

3

4

5

←1

WINTER SAFETY

←2

←8

←10

TRIGGER HOUSING 11→

9→

TRIGGER GUARD

←1

9A→

Figure 35. Trigger assembly - exploded view.

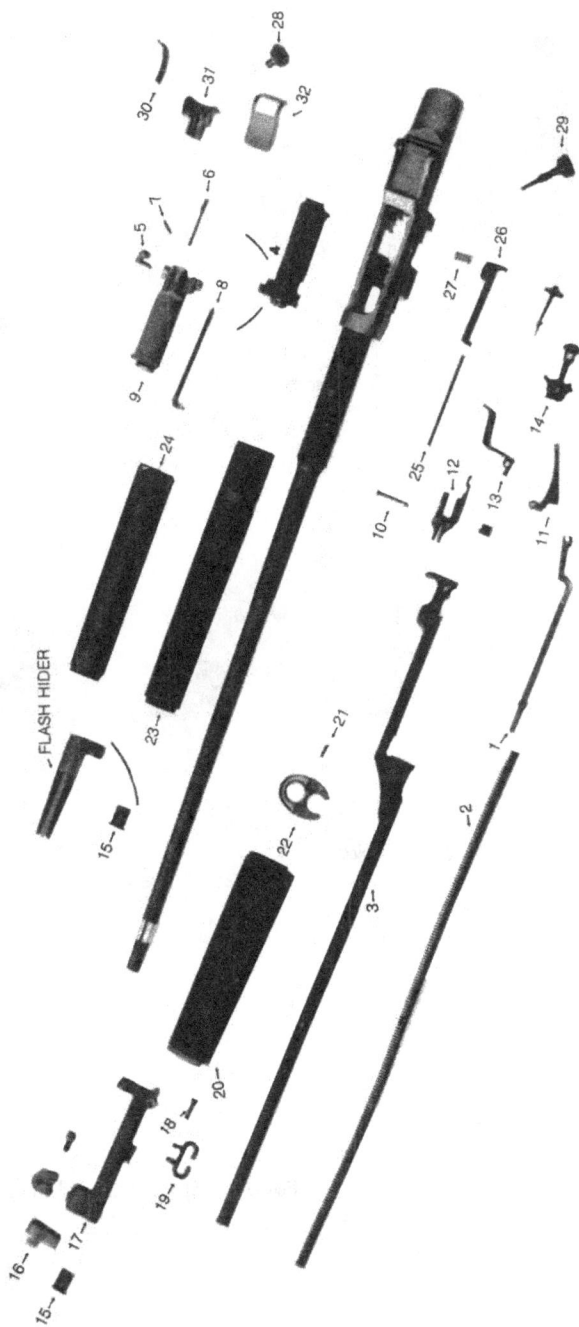

FLASH HIDER

Figure 36. Barrel and receiver group - exploded view.

53

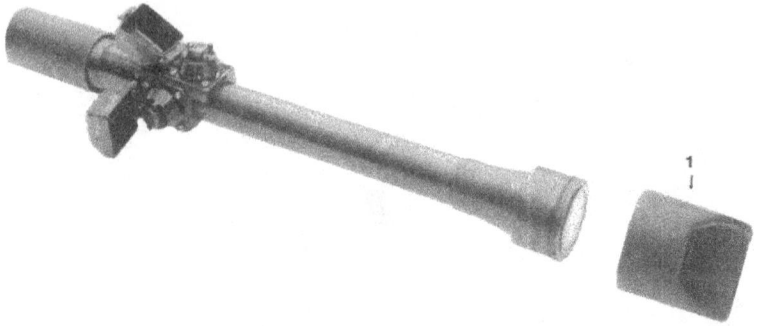

Figure 37. Telescope, M 84.

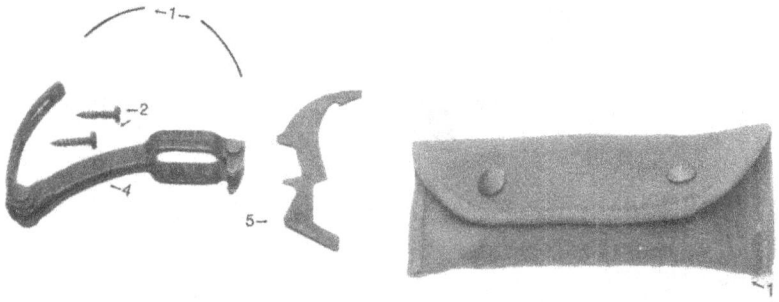

Figure 38. Winter trigger kit - exploded view.

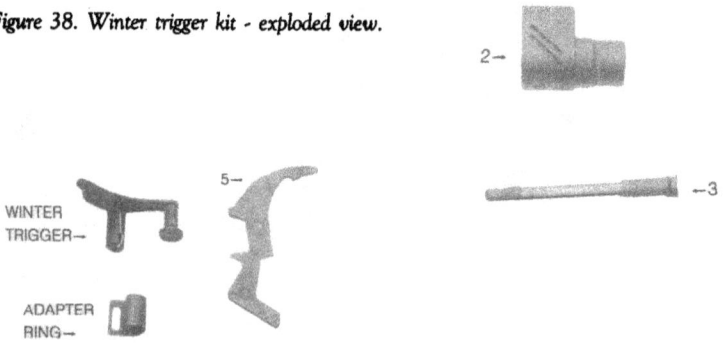

WINTER
TRIGGER→

ADAPTER
RING→

Figure 39. Winter trigger kit, type T-36.

Figure 40. Special tools and equipment.

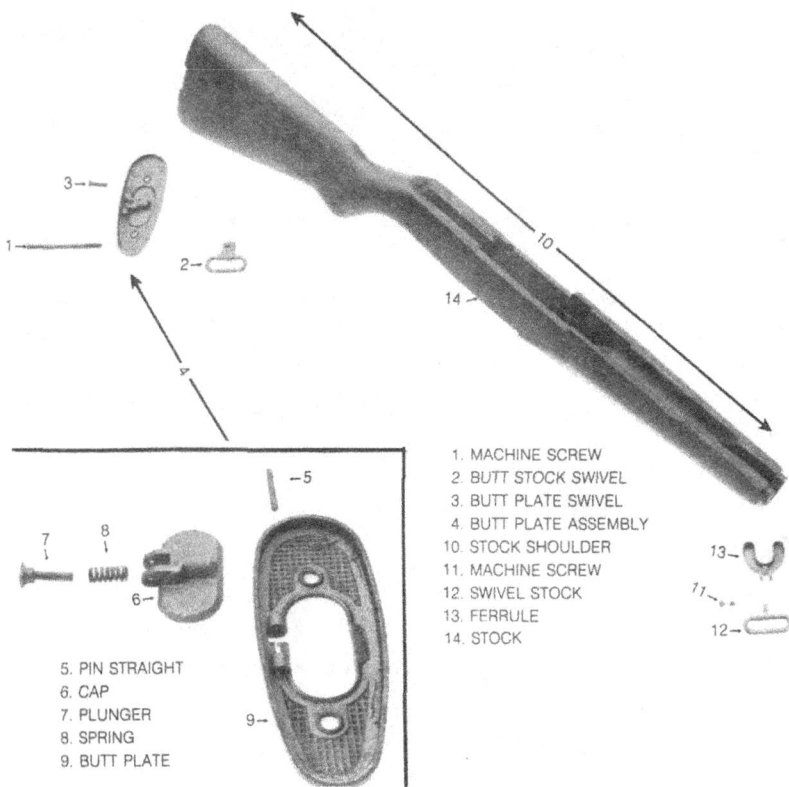

1. MACHINE SCREW
2. BUTT STOCK SWIVEL
3. BUTT PLATE SWIVEL
4. BUTT PLATE ASSEMBLY
10. STOCK SHOULDER
11. MACHINE SCREW
12. SWIVEL STOCK
13. FERRULE
14. STOCK

5. PIN STRAIGHT
6. CAP
7. PLUNGER
8. SPRING
9. BUTT PLATE

fig. 8
Figure 41. Stock group - exploded view.

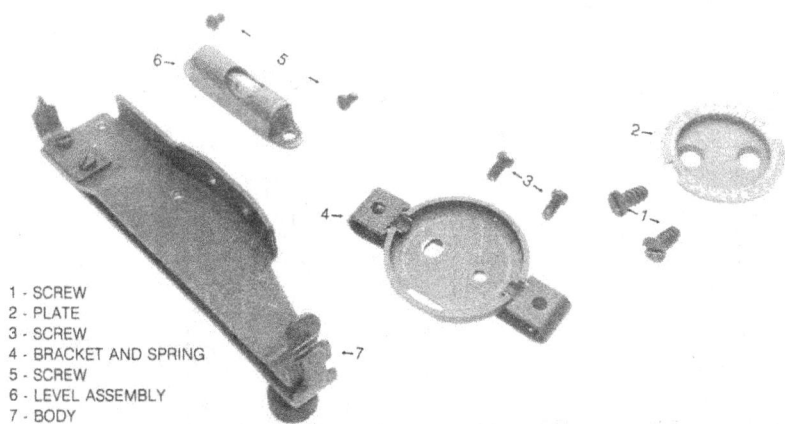

1 - SCREW
2 - PLATE
3 - SCREW
4 - BRACKET AND SPRING
5 - SCREW
6 - LEVEL ASSEMBLY
7 - BODY

Figure 42. Grenade launcher sight, M 15 - exploded view.

Figure 43. Gas cylinder cap screws, current type and old type.

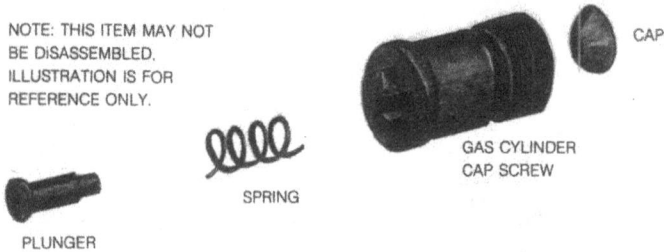

NOTE: THIS ITEM MAY NOT BE DISASSEMBLED. ILLUSTRATION IS FOR REFERENCE ONLY.

CAP

GAS CYLINDER CAP SCREW

SPRING

PLUNGER

Figure 44. Gas cylinder cap screw, current type — exploded view.

Appendix C
MAINTENANCE ALLOCATION CHART

Section I. INTRODUCTION

C-1. General

The maintenance allocation chart indicates specific maintenance operations performed at proper maintenance levels. Deviation from maintenance operations allocated in the chart is authorized only upon approval of the Commanding Officer.

Section II. Maintenance Allocation Chart for Rifle, Caliber .30, M1, M1C (Sniper's), and M1D (Sniper's)

(1) Group number	(2) Functional group	(3) Maintenance function											(4) Tool and equipment	(5) Remarks
		Inspect	Test	Service	Adjust	Align	Calibrate	Install	Replace	Repair	Overhaul	Rebuild		
1	TRIGGER HOUSING ASSEMBLY	C	—	C	—	—	—	C	—	O	D	—		
2	STOCK ASSEMBLY	C	—	C	—	—	—	C	—	F	D	—		
3	BARREL AND RECEIVER GROUP	C	—	C	—	—	—	—	—	O	D	—		
4	TELESCOPE, M84	C	—	C	—	—	—	—	H	—	D	—		
5	TELESCOPE, W/BRACKET	C	—	C	—	—	—	C	—	O	D	—		
6	MOUNT ASSEMBLY, M1C	C	—	C	—	—	—	F	—	F	D	—		
7	MOUNTING BRACKET ASSEMBLY M1D	C	—	C	—	—	—	F	—	F	D	—		
8	CHEEK PAD ASSEMBLY, M1C & M1D	C	—	C	—	—	—	F	F	—	D	—		

C-2. Maintenance Functions

The maintenance allocation chart designates overall responsibility for the maintenance function of an end item or assembly. Maintenance functions will be limited to and defined as follows:

INSPECT — To determine serviceability of an item by comparing its physical, mechanical, and electrical characteristics with established standards.

TEST — To verify serviceability and to detect electrical or mechanical failure by use of test equipment.

SERVICE — To clean, to preserve, to charge, and to add fuel, lubricants, cooling agents, and air.

ADJUST — To rectify to the extent necessary to bring into proper operating range.

ALIGN — To adjust specified variable elements of an item to bring to optimum performance.

CALIBRATE — To determine the corrections to be made in the readings of instruments or test equipment used in precise measurement. Consists of the comparison of two instruments, one of which is a certified standard of known accuracy, to detect and adjust any discrepancy in the accuracy of the instrument being compared with the certified standard.

INSTALL — To set up for use in an operational environment such as an emplacement, site, or vehicle.

REPLACE — To replace unserviceable items with serviceable assemblies, subassemblies, or part.

REPAIR — To restore an item to serviceable condition. This includes, but is not limited to, inspection, cleaning, preserving, adjusting, replacing, welding, riveting, and strengthening.

OVERHAUL — To restore an item to a completely serviceable condition as prescribed by maintenance serviceability standards using the Inspect and Repair Only as Necessary (IROAN) technique.

REBUILD — To restore an item to a standard as nearly as possible to original or new condition in appearance, performance, and life expectancy. This is accomplished through complete disassembly of the item, inspection of all parts or components, repair or replacement of worn or unserviceable elements (items) using original manufacturing tolerances and specifications, and subsequent reassembly of the item.

C-3. Explanation of Format

Purpose and use of the format are as follows:

a. Column 1, Group Number. Lists group numbers, to identify components and assemblies.

b. Column 2, Functional Group. Lists the noun names of groups and assemblies on which maintenance is authorized.

c. Column 3, Maintenance Functions. Lists the various categories of maintenance to be performed on the weapon.

d. Use of Codes. Explanation of the use of codes in maintenance function, column 3, is as follows:

Code	Explanation	Code	Explanation
C	. . . . Operator/crew		
O	. . . . Organizational	H	. . . . General Support
F	. . . . Direct Support	D	. . . . Depot

e. Column 4, Tools and Equipment. This column will be used to specify those tools and test equipment required to perform the designated function.

f. Column 5, Remarks. Self-explanatory.

Note: Columns not utilized are considered not applicable.

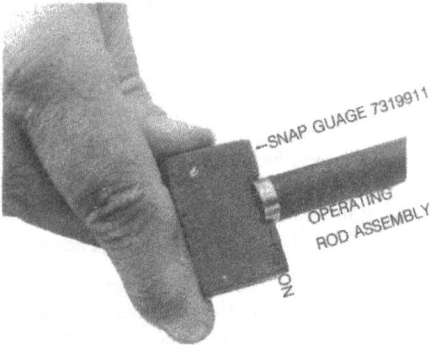

Figure 45. Gaging diameter of gas piston.

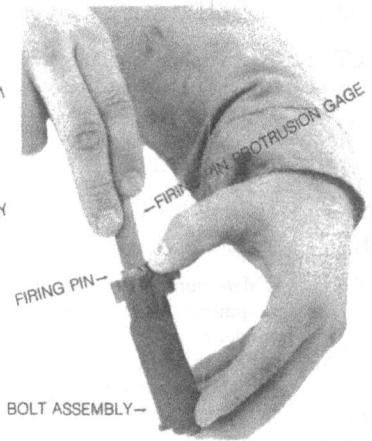

Figure 47. Gaging firing pin protrusion.

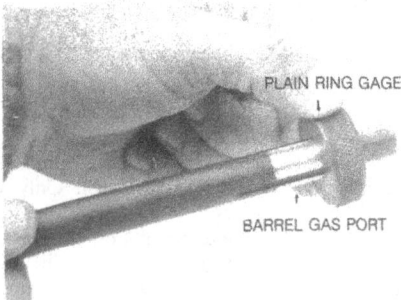

Figure 46. Gaging barrel diameter at gas port.

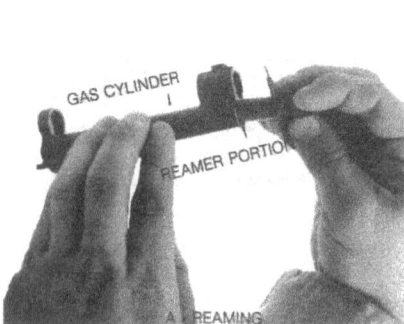

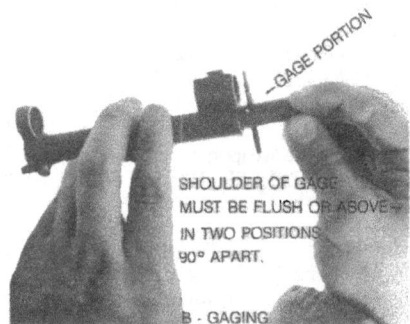

Figure 48. Reaming and gaging front interior of gas cylinder.

RECEIVER GAGE

GAGE HANDLE

TANG OF GAGE PLUG

GAGE PROPERLY POSITIONED IN RECEIVER

GAGE PLUG BELOW GAGING SURFACE

—GAGING SURFACE

PLUG POSITIONED
FLUSH OR BELOW

PLUG STEP POSITIONED
FLUSH OR BELOW

—IGNORE

—GAGING
SURFACE

CHECKING GAGE —
RECEIVER SERVICEABLE

CURRENT DESIGN

ORIGINAL DESIGN

SERVICEABLE

GAGE PLUG ABOVE GAGING SURFACE

GAGING SURFACE

PLUG PROTRUDES

PLUG STEP PROTRUDES

—IGNORE

—GAGING
SURFACE

CURRENT DESIGN

ORIGINAL DESIGN

CHECKING GAGE - RECEIVER UNSERVICEABLE

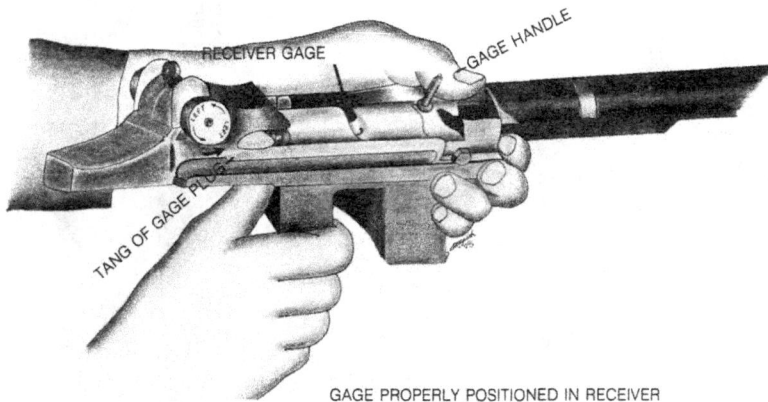

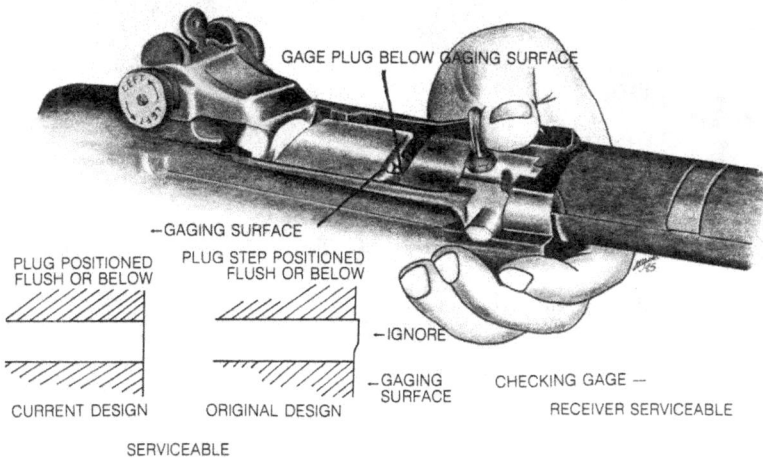

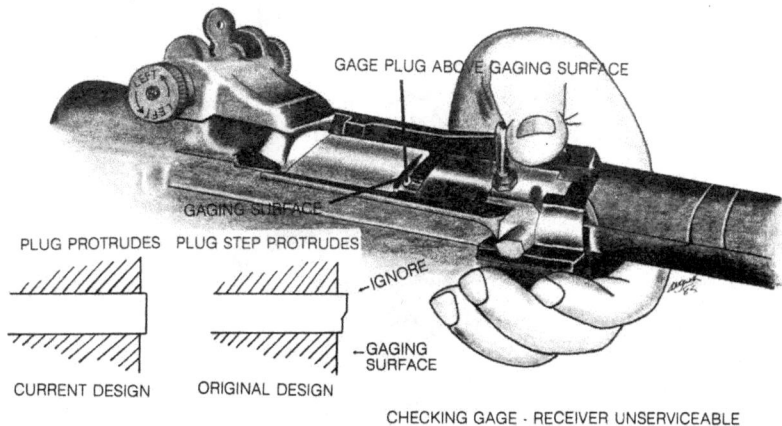

Figure 49. Gaging receiver.

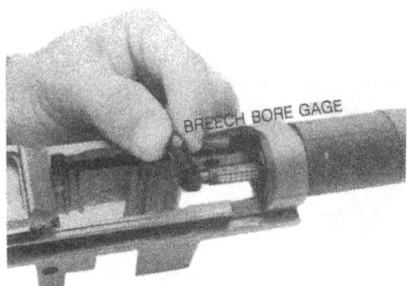

REJECT LIMIT 0.310

Figure 50. Gaging breech bore.

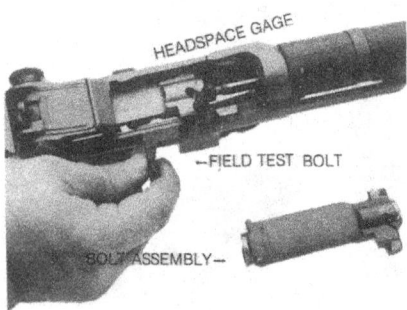

—FIELD TEST BOLT

BOLT ASSEMBLY—

Figure 51. Checking headspace.

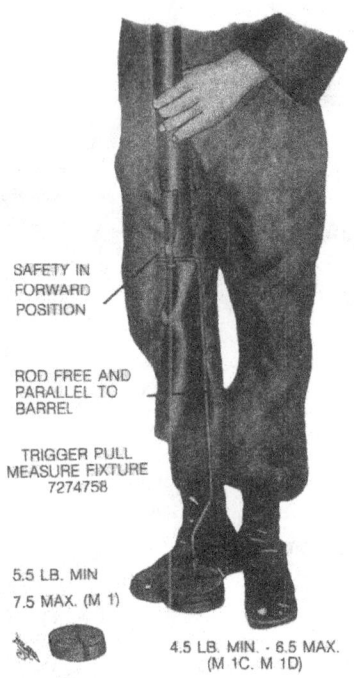

SAFETY IN FORWARD POSITION

ROD FREE AND PARALLEL TO BARREL

TRIGGER PULL MEASURE FIXTURE 7274758

5.5 LB. MIN
7.5 MAX. (M 1)

4.5 LB. MIN. - 6.5 MAX.
(M 1C. M 1D)

Figure 52. Checking trigger pull.

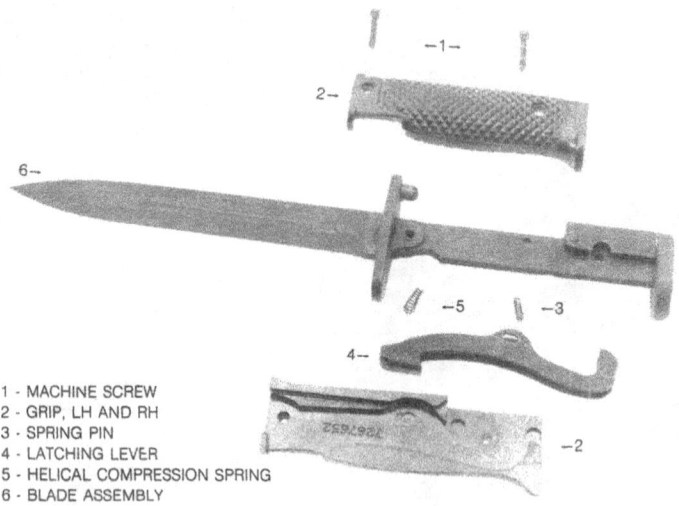

1 - MACHINE SCREW
2 - GRIP, LH AND RH
3 - SPRING PIN
4 - LATCHING LEVER
5 - HELICAL COMPRESSION SPRING
6 - BLADE ASSEMBLY

Figure 53. Bayonet knife, M 5A1 - exploded view.
NOTE: Bayonet knife M 5 differs internally, only, in position of helical spring.

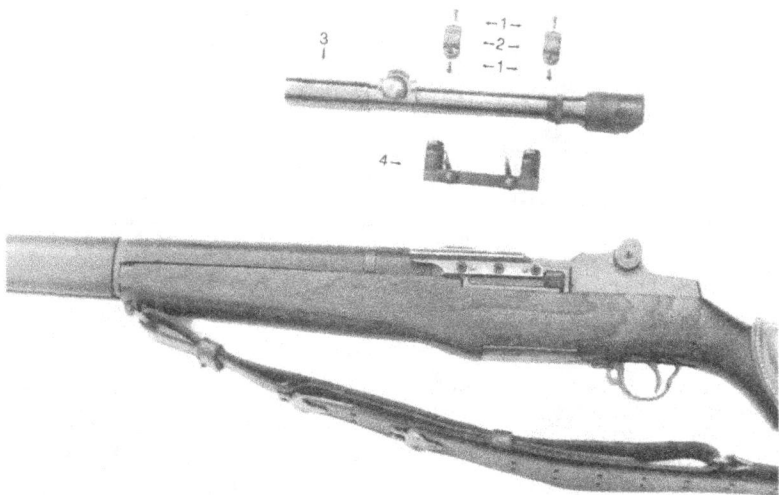

Figure 54. Telescope, M 82 and mount assembly (M 1C, Sniper only). From the collection of Bill Douglas.

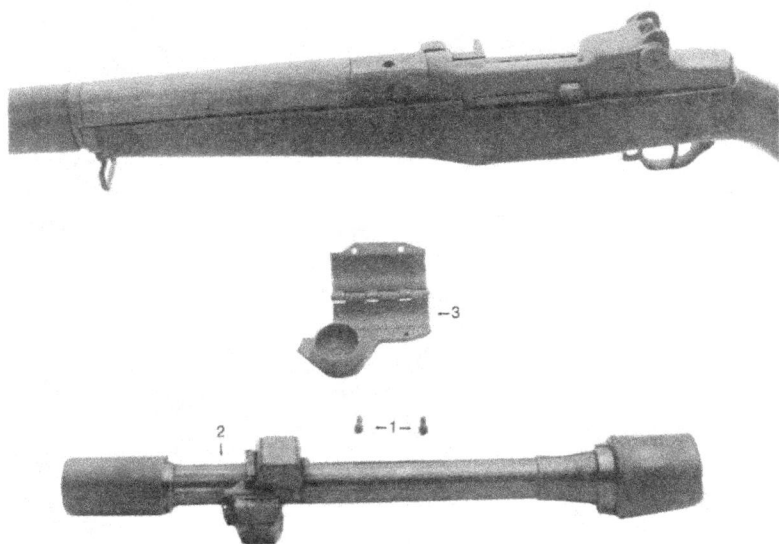

Figure 55. Telescope, M 84 and mount assembly (M 1D, Sniper only).

TOP PAD INSERT
MIDDLE INSERT

CHEEK BASE
AND PAD—

REMOVE/INSTALL CHEEK PAD.

LACE—
STOCK ASSEMBLY—

CHEEK PAD WITH LACE INSTALLED ON RIFLE, M 1C
OR M 1D.

DRILL NO. 31 (0.120)
HOLES 1 IN. DEEP

3.25 IN.

3.75 IN.

DRILLING HOLES IN STOCK ASSEMBLY FOR INSTALLING
WOOD SCREWS.

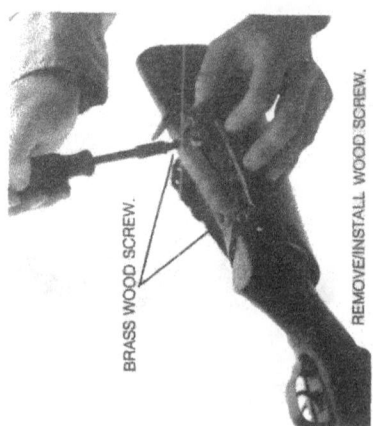

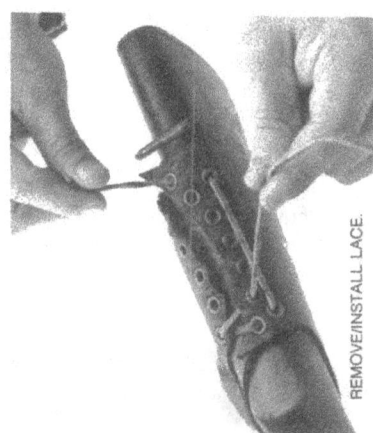

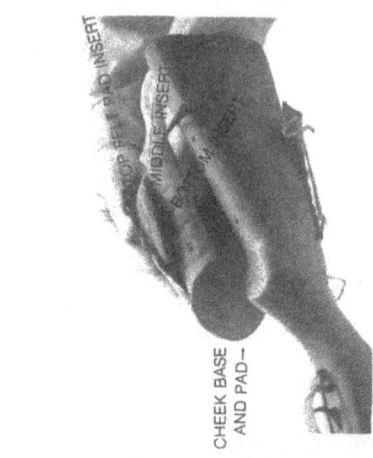

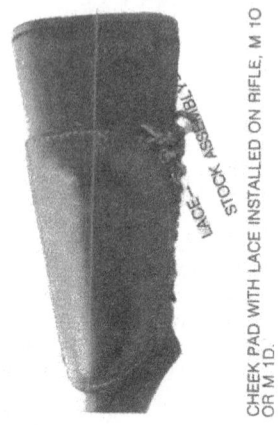

REMOVE/INSTALL LACE.

CHEEK PAD—

—SCREWS—

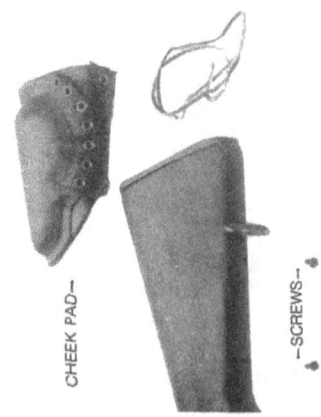

CHEEK PAD ASSEMBLY — EXPLODED VIEW
(M 1C (Sniper's) and M 1D (Sniper's) only).

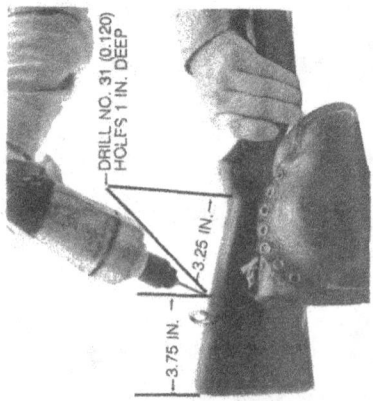

BRASS WOOD SCREW.

REMOVE/INSTALL WOOD SCREW.

Figure 56. Installation/removal of cheek pad assembly, Rifles M 1C and M 1D (Sniper's)

WINTER TRIGGER ASSEMBLY INSTALLED.

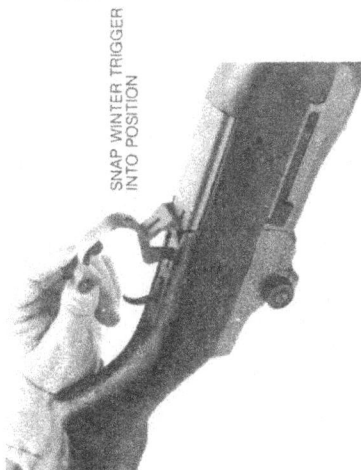

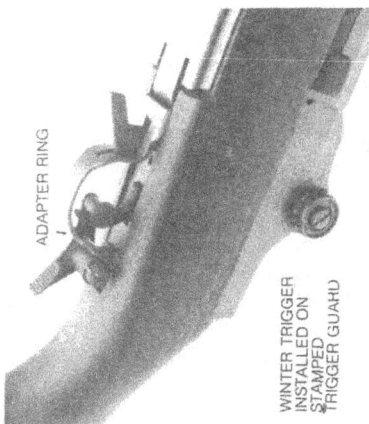

WINTER TRIGGER ASSEMBLY

LEVER

CAM

WINTER SAFETY

TRIGGER GUARD

WOOD SCREW

WASHER

ADAPTER RING

WINTER TRIGGER INSTALLED ON STAMPED TRIGGER GUARD

Figure 57. Installation of winter trigger bit.

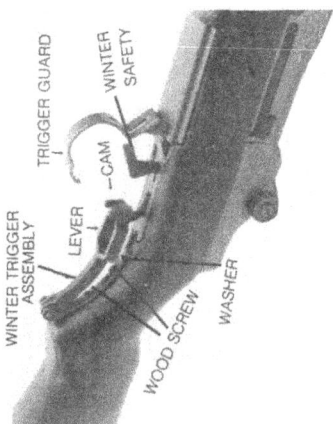

1/16 INCH DRILL

STOCK

DRILLING HOLES IN STOCK ASSEMBLY

CLOSE TRIGGER GUARD

LEVER

HINGE

REGULAR TRIGGER

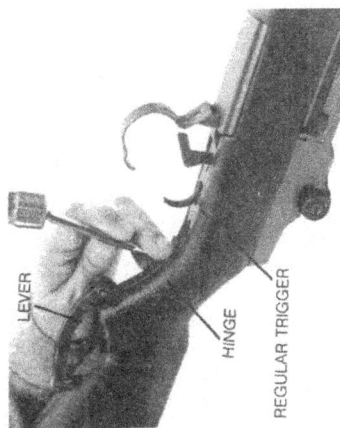

MARKING LOCATION FOR DRILLING HOLES

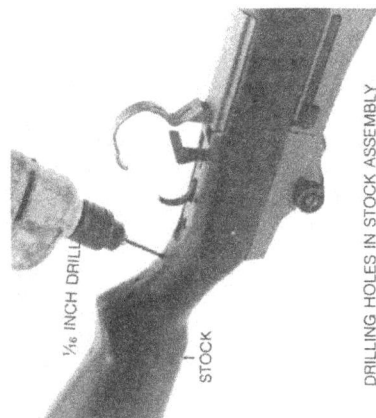

SNAP WINTER TRIGGER INTO POSITION

Figure 58. Installation of alternate winter trigger bit. Type T-36

By Order of the Secretary of the Army:

W.C. WESTMORELAND,
General, United States Army,
Chief of Staff.

Official:
KENNETH G. WICKHAM,
Major General, United States Army,
The Adjutant General.

DISTRIBUTION:
To be distributed in accordance with DA Form 12-40, (qty rqr block no. 128) Organizational maintenance requirements for Rifles, Caliber .30: M1, M1C (Sniper's) and M1D (Sniper's).

www.ingramcontent.com/pod-product-compliance
Lightning Source LLC
Chambersburg PA
CBHW070946210326
41520CB00021B/7078